T0142502

Gaming Media and Social Effects

Editor-in-chief

Henry Been-Lirn Duh, Hobart, Australia

Series editor

Anton Nijholt, Enschede, The Netherlands

More information about this series at http://www.springer.com/series/11864

Yiyu Cai · Sui Lin Goei · Wim Trooster
Editors

Simulation and Serious Games for Education

Springer

Editors
Yiyu Cai
School of Mechanical and Aerospace
  Engineering
Nanyang Technological University
Singapore
Singapore

Wim Trooster
Windesheim University of Applied Sciences
Zwolle, Overijssel
The Netherlands

Sui Lin Goei
Windesheim University of Applied Sciences
Zwolle, Overijssel
The Netherlands

ISSN 2197-9685   ISSN 2197-9693   (electronic)
Gaming Media and Social Effects
ISBN 978-981-10-9270-1   ISBN 978-981-10-0861-0   (eBook)
DOI 10.1007/978-981-10-0861-0

Printed on acid-free paper

This Springer imprint is published by Springer Nature
The registered company is Springer Science+Business Media Singapore Pte Ltd.

Foreword

In 2014, the book "Simulations, Serious Games and their Applications" by Yiyu Cai and Sui Lin Goei was published. That same year Windesheim University of Applied Sciences (The Netherlands) in cooperation with Nanyang Technological University (Singapore) organized the second "Asia-Europe Symposium on Simulations and Serious Games" in Zwolle, The Netherlands. A rich variety of seminars and workshops from Singapore, China, The Netherlands, and other European countries provided an excellent overview of new developments in the educational use of Serious Games and Simulations.

Every two years an Asia-Europe Symposium will be held with seminars and workshops designed to showcase the newest achievements in Serious Games and Simulation for educational purposes. It will be organized, alternating between Asia and Europe. This coming December month, the third Asia-Europe Symposium on Simulations and Serious Games will be held in Zhuhai, China in conjunction with the 2016 ACM SIGGRAPH International Conference on Virtual-Reality Continuum and Applications in Industries (VRCAI 2016), more information of the conference can be found at http://www.vrcai.org

All over the world in the field of education we see that pedagogical and didactical approaches of personalized learning are changing very fast. Simulations, Serious Games, Robotics, and Virtual Reality will inevitably be used more and more in our classrooms, to support the learning and teaching process.

I expect that this second book by Yiyu Cai, Sui Lin Goei and Wim Trooster will tell or show educators and researchers which new ideas, insights, information and experiments there are in the field of Serious Games and Simulation. I am also sure that it will inspire us to use these new educational methods that will enable us to improve education.

Harry Frantzen
Windesheim University of Applied Sciences
The Netherlands

Preface

This new book is a continuous effort in promoting simulations and serious games with a focus on educational application. The major part of this book is selected from the work presented at the 2014 Asia–Europe Symposium on Simulation and Serious Gaming held in Windesheim University of Applied Sciences, The Netherlands (Oct 1–2, 2014). All chapters in this book went through a peer-review process.

Three major domains of education using simulation and serious games are covered in the book including (1) Science, Technology, Engineering, and Mathematics (STEM) Education; (2) Humanity and Social Science Education; and (3) Special Needs Education.

For STEM Education, Wouter van Joolingen discusses science education in Chapter "A Germ for Young European Scientists: Drawing-Based Modelling"; Panpan Cai et al. address vocational training in Chapter "Simulation-Enabled Vocational Training for Heavy Crane Operations"; Hester Stubbé et al. present game based mathematic education in Chapter "Formative Evaluation of a Mathematics Game for Out-of-School Children in Sudan"; Diana Zwart et al. share vocational math teachers training in Chapter "Empowering Vocational Math Teachers by Using Digital Learning Material (DLM) with Workplace Assignments".

For Humanity and Social Science Education, Jaap van der Molen et al. present their game based learning in Chapter "The Odyssey Game"; and Wim Trooster et al. share game based English pronunciation learning in Chapter "The Effectiveness of the Game LINGO Online: A Serious Game for English Pronunciation".

For Special Needs Education, Yiyu Cai et al. discuss their research project in Chapter "The Virtual Pink Dolphins Project: An International Effort for Children with ASD in Special Needs Education"; Lennard Chua et al. share their emotional learning project in Chapter "ICT-Enabled Emotional Learning for Special Needs Education". Zengguo Ge and Li Fan present their social development project in Chapter "Social Development for Children with Autism Using Kinect Gesture Games: A Case Study in Suzhou Industrial Park Renai School"; and Yeli Feng and

Yiyu Cai describe a project for children with autism spectrum disorders in Chapter "A Gaze Tracking System for Children with Autism Spectrum Disorders".

Researchers and developers in simulation and serious games for educational uses will benefit from this book. Training professionals and educators can also benefit from this book to learn the possible application of simulation and serious games in various areas.

Singapore   Yiyu Cai
Zwolle, The Netherlands   Sui Lin Goei
Zwolle, The Netherlands   Wim Trooster

Contents

The Virtual Pink Dolphins Project: An International Effort for Children with ASD in Special Needs Education

Yiyu Cai, Ruby Chiew, Li Fan, Meng Kiam Kwek and Sui Lin Goei

Abstract The number of reported cases of Autism Spectrum Disorders (ASD) has increased rapidly in the recent years. Virtual Reality (VR) as a technology has been studied as an augmented intervention for children with ASD in special needs education and neuro-rehabilitation. This chapter will report an international effort by researchers, developers, and educators from Singapore, The Netherlands, and China to help children with ASD in their learning. VR technology is developed with an aim to create immersive learning environments for special needs schools. Educators and developers work closely to design learning content for children with ASD in their learning through interactive gaming. The partnership enables collaboration at different levels including research and development, sharing and exchanges, and so on. International symposia are organized under this international partnership to share knowledge and experience in special needs education. Exchange programs are also developed for school teachers and students through this international collaboration.

Keywords Special needs education · Simulation · Serious games · Virtual reality

Y. Cai (✉)
Nanyang Technological University, Singapore, Singapore
e-mail: myycai@ntu.edu.sg

R. Chiew
AWWA Special School, Singapore, Singapore

L. Fan
Suzhou Industrial Park RENAI School, Suzhou, China

M.K. Kwek
Underwater World Singapore, Singapore, Singapore

S.L. Goei
Windesheim University of Applied Science, Zwolle, The Netherlands

© Springer Science+Business Media Singapore 2017
Y. Cai et al. (eds.), *Simulation and Serious Games for Education*,
Gaming Media and Social Effects, DOI 10.1007/978-981-10-0861-0_1

1

1 Introduction

Children with Autism Spectrum Disorder (ASD) have impairments in social interaction and communication. They often have sensory processing difficulties and attention abnormalities predisposing them to process information differently and respond in unusual ways. Many people have conducted research regarding the ASD learning difficulties and behavioral challenges [1–4].

Virtual Reality (VR) technology has received good interest from the ASD research community for improving the learning of children with ASD. Austin et al. [5] investigated the feasibility of the use of VR hypnosis with two cases of ASD. Cai et al. [6] reported their design and development of a virtual dolphinarium for children with ASD.

This chapter will present an international effort to do research and development on VR technology for the purpose to improve children with ASD in their learning. Researchers, developers, and educators from Singapore, The Netherlands, and China tie up through this international collaboration. VR technology is developed with an aim to create immersive learning environments for special needs schools participating in the project. Educators and developers work closely to design learning content for children with ASD in their learning through interactive gaming. The partnership enables collaboration at different levels including joint research, visiting, sharing, and exchanges. International symposia are organized under this international partnership to share knowledge and experience in special needs education. Exchange programs are also developed for school teachers and students as part of the benefits from this international collaboration.

2 Participating Parties

Supported by The Temasek Trust Funded Singapore Millennium Foundation, the Virtual Pink Dolphins project is hosted by Nanyang Technological University (NTU) with various participating parties from Singapore, The Netherlands, and China.

2.1 Nanyang Technological University, Singapore

NTU is a young and research-intensive university in Singapore. In 2010, the Institute for Media Innovation at NTU funded a seed grant project to investigate the use of VR technology for children with ASD. In this pilot research, a 3D immersive room was set up and Virtual Pink Dolphins were prototyped (Fig. 1). As a multi-disciplinary research in NTU, the project attracted The Institute for Media Innovation, School of Mechanical and Aerospace Engineering, National Institute of

Fig. 1 The virtual dolphinarium with the immersive room at the Institute for Media Innovation, NTU

Education, School of Computer Engineering, and School of Art, Media and Design contributed in this initial phase. In 2013, the Virtual Pink Dolphins project received funding support from Singapore Millennium Foundation.

2.2 AWWA Special School, Singapore

Asia Women's Welfare Association (AWWA) School is a partner with this Virtual Pink Dolphins project. A simplified version of the Immersive Room was set up in the school (Fig. 2). AWWA School is the first special school in Singapore embarking VR technology to help children with ASD in their learning. Working closely with NTU's Institute for Media Innovation, AWWA School is the 2014 winner of the SPED Innovation Award by Singapore's Ministry of Education and National Council of Social Service.

2.3 RENAI School, China

Located in China–Singapore Suzhou Industrial Park, RENAI Special School is a partner school from China with the Virtual Pink Dolphins project. A simplified version of the Immersive Room was set up with RENAI (Fig. 3) providing a VR platform for children with ASD in their learning through gaming.

Fig. 2 Children with ASD at AWWA School play with Virtual Pink Dolphins in their own immersive classroom

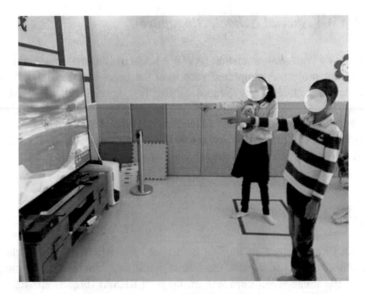

Fig. 3 Children with ASD at RENAI School play with Virtual Pink Dolphins in their own immersive classroom

Fig. 4 An Immersive Learning Corner was set up in the Lab 21 with the Windesheim University of Applied Sciences, The Netherlands

2.4 Windesheim University of Applied Sciences, The Netherlands

The Virtual Pink Dolphins project also has a partner in Europe. Windesheim University of Applied Sciences from The Netherlands has developed a cross-continental collaboration with Nanyang Technological University and AWWA School in Singapore, and RENAI School in China. An Immersive Learning Corner was set up in their Lab 21 at the University allowing researchers and educators to do research in SPED using VR technology (Fig. 4).

2.5 Underwater World Singapore

Underwater World Singapore (UWS) is located in the Sentosa Island. UWS has a tourism attraction with their pink dolphins, also known as Indo-pacific humpback dolphins. These dolphins provide visitors a unique entertaining experience through the popular dolphin show (Fig. 5a). UWS hosted the Pink Dolphin Encounter program (Fig. 5b) designed for children with ASD to interact with these very intelligent marine creatures. This program helped boost the self-confidence of these children when they physically encounter the pink dolphins in the lagoon [7].

The research team with the Virtual Pink Dolphins project works closely with the professional dolphin trainers from the Underwater World Singapore to understand the behaviors of the pink dolphins before the dolphin game is designed. They also had many communications during the game development.

Fig. 5 **a** Dolphin shows at UWS's new dolphin lagoon; **b** a child with ASD swimming with pink dolphins at UWS's old dolphin lagoon

3 International Collaboration

Several activities have been initiated with the Virtual Pink Dolphins project as international collaboration. This includes international conference organization, school visits and exchanges, joint publications, and research collaboration.

3.1 Asia–Europe Symposium on Simulations and Serious Games

On May 9, 2012, The First Asia–European Symposium on Simulations and Serious Games was held in Nanyang Technological University as part of the 25th International Conference of Computer Animation and Social Agents (CASA 2012). Selected presentations at the Symposium were invited as book chapters after necessary enhancements made and a book titled Simulations, Serious Games and Their Applications (Fig. 6a) was published by Springer in 2014 [8].

The Second Asia–Europe Symposium on Simulations and Serious Games was organized by The Windesheim University of Applied Sciences in Zwolle, the Netherlands from Oct 1–2, 2014. Partners with the Virtual Pink Dolphins project made presentations at the symposium (Fig. 6b). Nanyang Technological University, Windesheim University of Applied Sciences, AWWA School, and RENAI School made book chapter contributions in this current book published again by Springer.

(a)   **(b)**

Fig. 6 **a** The book—simulations, serious games and their applications published by Springer; **b** Presentation by RENAI School at the second Asia–Europe symposium on simulations and serious games

3.2 Schools Visits, Sharing, and Exchanges

Partners with the Virtual Pink Dolphins project have been actively involved in sharing and exchanges through various visiting and meeting arrangements.

On September 30, 2014, delegates from AWWA School and RENAI School visited De Ambelt, a Dutch SPED school, in Zwolle, the Netherlands. Three special schools exchanged their experiences during the meeting (Fig. 7a). Professors from Windesheim University of Applied Sciences visited AWWA School in Singapore on November 13, 2014 (Fig. 7b) and RENAI School in Suzhou, China on November 18, 2014 (Fig. 7c).

3.3 Students Internet Meetings

Apart from the visitings and meetings by researchers and educators under the partnership program, students from the partner schools have also been connected via Internet. Figure 8 shows the online chat between students from RENAI School in China and AWWA School in Singapore.

Fig. 7 Visitings and meetings under the partnership program. **a** Delegates from AWWA School and RENAI School visited De Ambelt in the Netherlands. **b** Professors from Windesheim University of Applied Sciences visited AWWA School in Singapore. **c** Professors from Windesheim University of Applied Sciences visited RENAI School in China

Fig. 8 Internet chat using SKYPE between RENAI students and AWWA students

Fig. 9 Signing of an MOU between Windesheim University of Applied Sciences and RENAI School

3.4 Memorandum of Understanding (MOU)

The Department of Teacher Education at Windesheim University of Applied Sciences, and the RENAI School with the China–Singapore Suzhou Industrial Park have entered an MOU for the partnership between the two parties (Fig. 9).

AWWA School and RENAI School have also signed an MOU for bilateral collaboration. The two schools will have more exchanges and visiting in the years to come.

4 Conclusion

The Virtual Pink Dolphins is a project initiated by the Institute for Media Innovation with the Nanyang Technological University, Singapore. Today, through international collaboration, researchers, developers, and educators are working closely in the project to design and develop SPED learning technology and learning contents. Also under the international partnership program, professors, principals, teachers, and students from different countries cross-continents are sharing and exchanging knowledge and experience of their common interests.

Acknowledgments This project is supported by The Temasek Trust Funded Singapore Millennium Foundation. Many people made contributions to this project. Special thanks go to Professor Nadia Thalmann, Associate Professor Jianmin Zheng, Professor Daniel Thalmann, Dr Indhumathi Chandrasekran, Mr Lihui Huang, Mr Nay Zin Tun, Miss Sandra Chan, Mr Zhengguo Ge, Ms Pauline Cheng, and Mr Norman Kee for their helps in one way or another.

References

1. Bloom, B.S., Engelhart, B.S., Purst, E.J., Hill, W.H., Krathwohl, D.R.: Taxonomy of Educational Objectives: the Classification of Educational Goals. Handbook I: Cognitive Domain. David McKay, New York, NY (1956)
2. Riggs, E.G., Gholar, C.R.: Strategies that Promote Student Engagement: Unleashing the Desire to Learn, 2nd edn. Corwin, Thousand Oaks, CA (2009)
3. Krathwohl, D.R., Bloom, B.S., Masia, B.B.: Taxonomy of Educational Objectives: The Classification of Educational Goals. Handbook II: Affective Domain. David McKay, New York, NY (1964)
4. Goldstein, E.B.: Sensation and Perception, 9th edn. Wadsworth, Belmont, CA (2013)
5. Austin, D.W., Abbott, J.M., Carbis, C.: The use of virtual reality hypnosis with two cases of autism spectrum disorder: A feasibility study. Contemp. Hypnosis **25**(2), 102–109 (2008)
6. Cai, Y., Chia, K., Thalmann, D., Kee, N., Zheng, J., Thalmann, N.: Design and development of a virtual dolphinarium for children with autism. IEEE Trans. Neural Syst. Rehabil. Eng. **21**(2), 208–217 (2013)

7. Watanabe, K., Lee, J.: Dolphin Encounter for Special Children (DESC): Proposal for Standard Procedures: Special Dolphin Program for Children with Autism, Down Syndrome, and Physically Disability. Underwater World Singapore, Singapore (2004)
8. Cai, Y., Goei, S.L.: Simulations, Serious Games and Their Applications. Springer, Berlin (2014)

A Germ for Young European Scientists: Drawing-Based Modelling

Wouter van Joolingen

Abstract An important movement in European science education is that learning should be inquiry-based and represents realistic scientific practice. The inquiry-based nature of science education is essential to interest more young people for a career in science and technology. Creating models is broadly seen as an essential part of those scientific practices. Dynamic models play a central role in science as a main vehicle to express and evaluate our understanding of complex systems. Therefore, the ability to reason with and about models and to create models of dynamic systems is an important higher order thinking skill and as a means to foster the development of scientific attitudes. In teaching children how to model, the choice for model representation is important. Representations can vary from mathematical formula, programming languages and diagrammatic representations. This chapter will present modelling based on drawings, and the SimSketch software with which children can create dynamic, multi-agent models. By representing systems in drawings, assigning behaviour to elements of the drawing and simulate the resulting model, children can express and test their ideas about natural and artificial systems. The chapter discusses conceptual and technical issues related to SimSketch as well as studies in which children have used SimSketch to represent systems such as the solar system, traffic and the spreading of diseases. The role of this approach will be discussed in the context of developments in European educational research.

Keywords Simulations · Modeling · Inquiry learning · Scientific literacy

W. van Joolingen (✉)
Freudenthal Institute for Science and Mathematics Education,
Utrecht University, Utrecht, The Netherlands
e-mail: w.r.vanjoolingen@uu.nl

© Springer Science+Business Media Singapore 2017
Y. Cai et al. (eds.), *Simulation and Serious Games for Education*,
Gaming Media and Social Effects, DOI 10.1007/978-981-10-0861-0_2

13

1 Introduction: The Image of Science in Science Education

In Europe, there is an increasing need for students choosing studies and careers in science and technology. Over the years, interest in science has decreased and numbers of students enrolling in scientific and technological studies have been going down. Although in recent years this downfall seems to have come to an end, it is important that the educational system raises interest, skill and motivation for science and technology.

Apart from the need for well-trained scientists, science education also needs to educate for the role in science as part of citizenship education. People need a basic level of knowledge about science in order to function in society, for instance to make choices regarding socio-scientific issues such as vaccination and genetic testing. According to Boerwinkel et al. [1], the scientific knowledge and skills that young people should acquire are

- knowledge and skills regarding concepts in science and mathematics;
- knowledge about the nature of science and mathematics, which includes methodology and foundations; and
- insights into norms and values, both personal and societal.

Together these concepts constitute a basis for empowering citizens in a knowledge society [2]. For instance, for deciding on whether or not to vaccinate their children, people need to know what vaccination is (concept knowledge), know about the value of the claims in favour of and against vaccination (nature of science), have insight into the consequences for themselves and others and associated personal and societal norms and values. Scientific citizenship relates knowledge and insights to a perspective on action that is related to function as a citizen. In this case, the action is making a decision about vaccination. We can therefore define scientific citizenship as the set of knowledge, skills, attitudes and values that contribute to a person's actions as a citizen.

For education in science this means a focus broader than teaching and learning the scientific concepts, by studying not only questions such as *what do we know?* but also *how do we know that?* and *why is it important to know?*

According to leading experts, science education in Europe fails both in making science attractive to young people and in educating science for citizenship. A crucial problem is that science is often taught as a collection of facts and skills to solve problems. For instance, Osborne and Dillon stated in 2008 [3]

In the past two decades, a consensus has emerged that science should be a compulsory school subject. However, whilst there is agreement that an education in science is important for all school students, there has been little debate about its nature and structure. Rather, curricula have simply evolved from pre-existing forms. Predominantly these curricula have been determined by scientists who perceive school science as a basic preparation for a science degree – in short a route into science. Such curricula focus on the foundational knowledge of the three sciences – biology, chemistry and physics. However, our contention is that such an education does not meet the needs of the majority of students who require a broad overview of the major ideas that science offers, how it produces reliable knowledge and the limits to certainty. Second, both the content and pedagogy associated with such curricula are increasingly failing to engage young people with the further study of science. Indeed, there is a strong negative correlation between students' interest in science and their achievement in science tests.

The negative correlation reported by Osborne and Dillon is worrying. Apparently science education is incapable of providing an image of science that is representative of actual scientific practice. Images of scientists as either nerdy, unworldly, or even "evil scientists" are prevailing in popular literature. Bonner [4] states that popular understanding of science suggests that scientific work does not grip the emotions, that it is coldly logical, and that it is not as creative as the actions of the artist. Yet, Bonner argues that much of scientific work is a combination of logical thinking and creative actions such as choosing a subject, finding questions, going against common beliefs, finding solutions, including unorthodox ones, etc. The problem with the "cold, logical" view of science is that it implies to many children that a scientific career is out of their reach, that it is uncreative and perhaps just plain "boring". As a result, talented girls and boys may be diverted from a route toward a career in science.

In the current chapter, an approach to science education for young children will be presented and explored in which young students engage in authentic scientific practice by creating scientific models and simulations. As such this approach is not new. The approach of modelling in education goes back to the 1980s in which time Jon Ogborn pioneered with this approach [5]. Recently, Louca and Zacharia [6] have reviewed the various approaches to modelling in education. We will introduce the basic principles of modelling in science and science education here and argue for the introduction of a more visual and qualitative way of modelling to address younger children. This will be followed by an introduction of SimSketch, our drawing-based modelling tool. The tool will be described as well as some experiences with the tool in educational practice.

2 Models in Science and Science Education

In introducing scientific practice to the practice of teaching science, much attention is given to the inquiry cycle [7, 8]. Whereas the inquiry cycle represents relevant processes in scientific inquiry, according to Windschitl and colleagues, a strict interpretation of this cycle may lead to the conception of inquiry as mechanically performing a fixed sequence of steps to test hypothesis, instead of a creative search for models and theories that allow us to form a coherent picture of the natural world [9]. In other words, there is not a single scientific method, but a range of activities that lead to the construction of scientific models.

The philosopher Giere [10] investigated the concept of models as objects that are created by scientists and form a central point in scientific discourse. In this line of reasoning, models are not meant to be a one-to-one representation of reality, but the relation between the model and reality serves a particular purpose in understanding particular features of reality. This has profound consequences for how we see the development of scientific knowledge. The purpose of science is not to discover "laws" that exist in some independent reality, but instead to create models to understand the things we observe. The same models can then help us to predict new observations and be used in the development of new solutions to societal problems. So, instead of being a discipline in which reality is "uncovered" using a fixed method of inductive and deductive steps, science is a *constructive* activity in which models of phenomena are constructed for particular purposes. This also means that multiple models can exist of the same phenomenon, each serving its own particular purpose. In modern science, models are often realised computationally. Astronomers, biologists, chemists and physicists alike use powerful computing techniques to simulate the models they create and use the computational results for predicting states of the world based on their models. For instance, astronomers use advanced simulation models to study the formation of stars and galaxies and to understand the structure of the universe as a whole and biochemists study the working of enzymes in cells using simulations. In science education also, computational modelling using systems such as STELLA [11], NetLogo [12, 13] and StageCast [14] have been designed so that students in secondary education can create models and become acquainted with the principles of scientific computational modelling.

In creating and understanding scientific models, visual representations play a key role in almost all areas of science [15, 16]. Visual representations are used to organize and constrain thought, relate concepts, visualize processes, abstract data and more [17]. In model building, these representations play a role both in *constructing* the model and *evaluating* it, using external data or other means. Especially in the constructive and creative phase, *drawing* is a very useful means of constructing visual representations: drawings allow the construction of representations without a priori constraints and rules of symbolic expression. This helps scientists and students alike to embed creative ideas into their models.

3 SimSketch: Creating Drawing-Based Models

This section introduces SimSketch. A more elaborate introduction is given elsewhere [18, 19], therefore only the main features will be highlighted.

3.1 Basic Design and Principles

A model in SimSketch always starts with a drawing of a system or process. For instance a drawing can depict the way energy flows between the sun and the earth [18] or the objects that compose of a mechanical system. The key idea behind

SimSketch is to combine such drawings with a modelling engine, so that the drawings can not only be used to show static structures in the learners' models, but can also become animations visualizing dynamic properties of the model.

In order to do so, the drawing is split into separate objects and each of the objects is assigned a behaviour. A clustering agent supports users in performing this splitting into objects. By guessing which drawing strokes belong together, based on spatial distribution and order of drawing, the clustering agents suggest a division in objects. Users can overrule this division if needed by drawing a grouping stroke around strokes that belong together.

The next step in SimSketch modelling is assigning behaviour to the objects. This behaviour can be an object's independent motion or an interaction with another object. For example, the GO behaviour specifies an object's independent motion in a specified direction, whereas the CIRCLE behaviour specifies that the object moves in a circular orbit around another object. Behaviours can be combined and can interact. For instance, if an object is assigned the CIRCLE behaviour and the object it is to circle around is moving, the circling object will move along too.

Interactive behaviours specify how objects respond to each other. Such behaviours include attractive and repulsive relations between objects. Also objects can "kill" or "eat" other objects. Finally, SimSketch offers behaviours for reproduction and cloning of objects.

After specifying behaviours, users can *run* the model, which creates an animated copy of their drawing in which the objects move according to the behaviours specified. Learners can zoom, speed up, or slow down the simulation and can have the simulation draw traces of the moving objects.

The design of SimSketch is aimed to support essential reasoning processes such as identifying model components, their properties and behaviour in an intuitive way. The object-based nature of the models allows learners to specify one object at a time in a relatively simple way, whereas complex behaviour may result from the combination of object and behaviours. SimSketch focuses on domains in which modelling results are best represented through animations, displaying qualitative aspects of complex behaviour. Two of these domains are elaborated below.

3.2   Modelling the Solar System

The first model investigated using SimSketch is that of the solar system. This model represents the movements of the planets around the sun. The drawing that forms the basis of the model displays the sun, some planets and, at least, the moon. Each of the objects, planets, moon and sun is labelled, as displayed in Fig. 1a. All planets receive the CIRCLE behaviour that instructs them to move around the sun. The moon also gets the CIRCLE behaviour with the instruction to move around the earth.

In this way, a *phenomenological* model of the solar system is created. The fact that the planetary orbits are in fact ellipses is neglected and no cause or explanation for the circular movement is given in this model. Also, no collision detection occurs

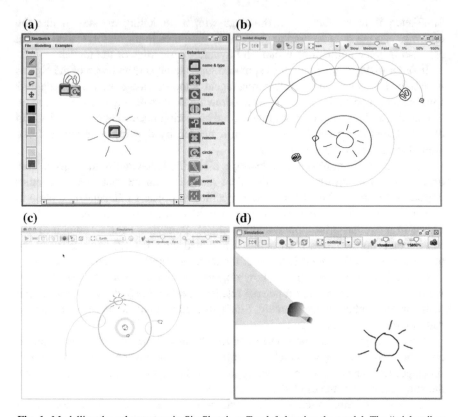

Fig. 1 Modelling the solar system in SimSketch. **a** *Top left* drawing the model. The "stickers" on the drawn elements specify the behaviour of each of the elements. **b** *Top right* simulating the model. Each of the moving planets leaves a trace as they move over the screen. **c** *Bottom left* by centering the earth instead of the sun the loops in the trajectory of Mars can be used to understand its retrograde movement. **d** *Bottom right* SimSketch allows indicating elements as light sources and light blockers, allowing to show the occurrence of eclipses

in this model—planets follow their orbits as specified. Still the model can be used to reveal several interesting properties of the solar system.

The first is the shape of the trajectories of satellites such as the moon. As is visible from Fig. 1b, the moon displays a cycloid orbit as seen from the sun. Moreover, by centering one of the planets, it can be made clear that the relative motion of the earth and the other planets may give rise to certain phenomena on the earthly sky. For instance, outer planets (those outside earth's orbit) display retrograde movement. If the position of—for instance—Mars on the sky is followed over the year, one notices that it will move from right to left most of the time, but in certain periods it will move in the opposite direction, the so-called retrograde movement. During retrograde, Mars also shines brighter at the evening sky. By fixing the position of the earth in the simulation and displaying Mars' orbit one can see the loops in Mars' orbit relative to the earth that explain this phenomenon as can be seen from Fig. 1c.

By adding light rays and shadows to the simulation, SimSketch can also explain the occurrence of solar and lunar eclipses (Fig. 1d). Here a limitation of the 2D nature of the model becomes apparent: explaining solar eclipses in this way should lead to an eclipse every month. In order to explain that this is not the case, a 3D perspective is necessary. Nevertheless, these examples show that the phenomenological model of the solar system that can be constructed in SimSketch has the power to explain relatively complex phenomena.

3.3 Modelling Multi-agent Systems: Traffic

As an illustration of multi-agent modelling we introduce traffic, and more specifically the spontaneous formation of congestions without an apparent cause. In cases of high traffic density, sometimes congestions occur spontaneously, without the presence of a clear cause, such as an accident or crossing.

In modelling such a system, two aspects are important in the model. One is that, even if all drivers intend to drive at the same speed, it is impossible to maintain *exactly* the same speed for all. The second is that drivers brake to avoid collisions with the car in front of them.

In SimSketch a traffic system can be constructed by drawing objects representing cars and an object that represents a circular road. Cars receive the behaviour to drive along the road with a give speed. In order to avoid having to draw many cars, a *factory* behaviour is associated to a single car object. This factory turns the car into a prototype of car objects that will be produced by the simulation. Apart from saving on drawing time, this also allows for experimentation with traffic density, by adjusting the number of cars that should be produced.

Figure 2 displays the model. Figure 2a displays the model as drawn, Fig. 2b shows the resulting simulation. Without extra measures, the simulation would result

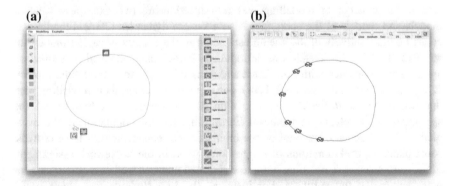

Fig. 2 Modelling traffic and congestions in SimSketch. **a** *Left* the drawn model, with as special element the factory that produces a specific number of cars. **b** *Right* the simulation with the cars moving around the drawn road displaying the spontaneous appearance of congestions

in cars moving around the road without any congestion, all with exactly the same speed. In order to display realistic behaviour, the factory can be adapted in such a way that a small randomization on the parameters (in this case speed) is applied to the objects produced. This has as result that the cars have slightly different speeds, and hence will collide. A second measure, making the cars adjust their speed to objects in front of them provides the observed behaviour.

Again, this model is phenomenological. There is no explanation of why the cars behave the way they do. This is both a limitation and strength of the model. The limitation relates to the explanatory power of the model, the strength is that it shows that with only a small number of assumptions, a rather complex phenomenon can be understood using the model. In designing learning activities around this model, such strengths and weaknesses should be addressed.

4 Experiences with SimSketch

SimSketch has been used in educational practice, within and outside school contexts. In this section, three practical uses of SimSketch are presented to display the range of application of the drawing-based modelling approach.

4.1 Master Classes

Many schools in the Netherlands nowadays sustain honours programs for their successful students. Within the context of these honours programs, the University of Twente offered master classes for eighth grade honours students. Six male eighth grade students aged 14 participated in a master class on modelling. The topic of the master class was scientific modelling and consisted of three 3-h sessions. In the first session the concept of modelling was introduced using two examples: weather models and models of the economy. Also a system dynamics [20, 21] model was created collaboratively of income and spending in the context of saving money for an iPad. Finally SimSketch was introduced. Participants received a homework assignment to practice with SimSketch and think of a topic for a modelling project. In the second session (one week later), participants presented their modelling plan that was discussed in the whole group. Chosen topics were traffic (two times), the spreading of the plague in medieval Europe, the python population in the everglades, the influence of alcohol on the brain and the management of large crowds. Most plans were elaborations of suggestions given in the homework assignment. Plans were discussed, resulting in more elaborated plans as well as requirements for new features in the software. For instance, the two students that chose traffic requested a behaviour that generated multiple instances of drawn objects that resulted in the factory behaviour (see Sect. 3.3). These new features were implemented in the days following the session. In the final session, students completed their models using the new features, and presented them to the group. In the presentations they were asked to justify the way the model had been built and why it could be a good model of reality.

Prior to the first session and at the end of the third session, participants filled in a questionnaire on the nature of models and their own stance toward science. The nature of models questionnaire consisted of open questions, for instance on how models should be tested. The opinion about science was measured using 4-point Likert scale questions, such as "I like to discover new things". The final questionnaire also contained questions for evaluating the software using a 5-point Likert scale, containing questions such as: "I would like to work with SimSketch again" and "SimSketch is easy to learn". Moreover the models created by the learners were collected in their increasing versions. During all three session observational notes were taken.

All participants created models that could be simulated and for which they indicated that they represented their ideas. These simulations showed behaviour that expressed at least one core idea, for instance the way populations fluctuate in a predator-prey system. One participant decided not to use SimSketch but instead used a demo version of a professional traffic simulator to show the effect of road works on traffic flow. As the teachers of the master class found it more important that students caught the purpose of modelling than to promote SimSketch they allowed this student to use this software that he had downloaded from the Internet.

On the questionnaire that evaluated SimSketch, participants evaluated SimSketch positively (mean value 3.81 on a five point scale, with 3 as neutral value).

In participants' views on science a remarkable change occurred. In the pretest the main view on models was that of scale models, which was mentioned by five of the six participants. Also five participants stressed that models should be as precise as possible to be a representation of reality and that not everything could be modelled. No participant mentioned models for reasoning or simulation. In the post-test, participants all mentioned reasoning and simulation, and stressed that almost everything can be subject of modelling. Also, they mentioned modelling of *ideas,* and that there can be multiple models of reality depending on the purpose of the model.

Despite the fact that the number of participants was small, and these honours students were certainly not average students, the study does reveal interesting aspects of model-based reasoning in teaching. The first was the ability of students to develop their own topic and a coherent model of that topic. They could do this without programming experience. SimSketch proved to be easy to learn, which was confirmed by the evaluation questions. Although the students initially did follow suggestions, after that all students had gathered documentation and related that to model elements. All students placed their model in a context—that was not given. For instance the predator-prey system was placed in the context of the python plague in the everglades and crowd management was placed in the context of a recent item in the news on panic in a big crowd.

Due to the small group, we had the opportunity to adapt the system to the needs in their individual model proposals. This contributed to the development of the

SimSketch system in such a way that the new features were developed on demand in the models created by the learners. Although not scalable, this method proves valuable for the development of advanced educational software and contributed to the engagement of the participants—as they could and did suggest elements of the new version of the program.

4.2 Solar and Lunar Eclipses

In a study to investigate whether young children (from age 7) are able to use SimSketch we designed a modelling task on the solar system. The main goal was to understand and explain the occurrence of solar and lunar eclipses. This task was offered to children visiting the NEMO science center in Amsterdam. They used SimSketch to create models of the solar system and were specifically asked to create situations in which solar and lunar eclipses occur.

247 participants (127 girls and 121 boys) in the age range of 7–15 participated. Participants filled in an 8-item multiple-choice test on their knowledge of the solar system that was specifically designed for this task. Questions included "What composes the solar system?" and "What happens at a solar eclipse?". They then received a short SimSketch tutorial. After that they created a model drawing of the solar system.

In the instruction they were asked to create an animated view of the solar system, including the light and shadow behaviours as depicted in Fig. 1d, and to stop and save the simulation at the moment a solar or lunar eclipse would occur. Participants filled in the domain knowledge post-test. Drawings were scored on the presence of essential elements in the representation of the solar system and relations between them. Data were analyzed to search for relations between age, gender and prior knowledge score, drawing score and post-test score. The main question was whether students would gain knowledge from this short modelling activity, whether the quality of the drawing contributed to their knowledge gain, and whether such results were dependent on age and gender.

Detailed results of this study are published elsewhere [22]. Here we summarize the main outcomes. On average, students worked on the task for approximately 40 min, including pre- and post-tests. This was the maximum that is realistic in the museum context. For instance, the parents of 15 of the children wanted to move on before completion of the task by their children. For these children the dataset was incomplete. These data were excluded from analysis.

Because no explicit instruction about the domain was given in this study, any knowledge gain between pre- and post-test should be attributed to the activity on the task performed by the children. Therefore, a large knowledge gain was not to be expected. The main scores on pretest and post-test are depicted in Fig. 3. Analysis of the pre- and post-test data revealed that there was a small but significant increase in score, from pre- to post-test. Detailed analysis showed that this increase could mainly be ascribed to the young students, between 7 and 9 years old. Older students

Fig. 3 Pretest and post-test scores, split for age groups and gender, indicated by ♂ and ♀. Maximum score is 8

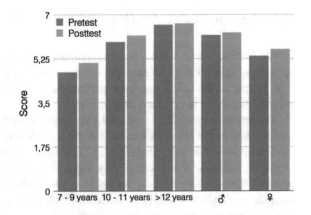

scored higher on both pre- and post-test, but gained less than the youngest group, probably due to a ceiling effect. Also it was found that girls gained significantly between pre- and post-test whereas boys did not. To investigate whether the drawing contributes to knowledge acquisition, we computed a partial correlation between post-test score and model score with pretest score partialled out. This correlation turned out to be significant (r (222) = 0.186, p = 0.005), indicating an influence of the drawing process on knowledge acquisition by the participants. Better drawings were associated with higher gains on the post-test. Further analysis of the data, using structural equation modelling showed that models with and without the quality of the drawing included fitted almost equally well on the data. Therefore we should see the relation between the quality of the model, as expressed in the number of elements included, with the learning gain as undecided and an issue for further study.

The results from the motivational scales and the participants' responses on the open question indicate the potential of the approach to motivate children to learn by engaging them into drawing-based modelling. Scores on how participants evaluated SimSketch, the perceived competence (the amount to which participants thought they could use SimSketch for the task) and valuing (the extent to which they thought that the task as a whole was interesting and attractive) were above average. Competency scored 2.98 and valuing scored 2.91, both on a scale from 1 to 4 with 2.5 as neutral value.

These results show that young children are capable of creating drawing-based models. Their model scores—in the range of 6.5–8.5 (max. 14)—indicate that they have included the major essential elements in the drawing, but omitted relatively unimportant details. Children from 12 years old and above display enough understanding of the solar system to create accurate models. Knowledge gains can be seen in younger groups, especially for girls. Summarizing, we see that drawing-based modelling is a feasible approach to learning scientific topics, within reach of even young children, and that it potentially contributes to the resolution of misconceptions in astronomy [23].

4.3 Lesson Design for Drawing-Based Modelling

In this section, we describe a first implementation of SimSketch in regular lessons in primary education as an example of how a modelling lesson may be designed. The lesson as described has been performed multiple times in sixth grade of a primary school in the Netherlands.

The lesson focuses on the principle of modelling and builds on the traffic model as described in the previous section. After a short introduction on the basic ideas of modelling and the role models play in science (using weather models as an example) students were asked if they knew situations where congestions seem to appear without apparent cause, for instance when driving with their parents. After a brief discussion, a short movie was shown in which cars drive on a circular road and in which spontaneous congestions occur.

Students were asked to think of possible causes of such traffic jams. As part of this process, they were asked to walk in a circle through the classroom while maintaining a constant speed (see Fig. 4). This experience was discussed. The main question in the discussion was whether students felt they could move freely and if not what caused any hindrance. Many students mentioned causes such as reaction time, inattentive drivers, etc. as causes of the traffic jam. In further and deeper discussion of these causes the teacher tries to reach consensus on two factors: cars

Fig. 4 Children playing out the traffic situation by walking around and assuming the role of cars

never drive at exactly the same speed and drivers want to avoid collision. Factors such as reaction time and braking are related to the variables speed and the tendency to avoid collisions.

The teacher then developed the model by drawing on an interactive whiteboard, asking students about what the model elements should be. After doing this, the teacher drew the elements on the board and explored the influence of the various behaviours assigned, such as the variation in speed.

When the model showed the behaviour that was observed in the movie, a class discussion followed in which the essential properties of the model were discussed. Three questions are central in this discussion: "Does the model explain the phenomenon?", "How does the model explain the phenomenon?" and "What are the limitations of the model?".

These three questions are essential in understanding the functions of this particular model and the role models play in understanding scientific phenomena. Therefore these three questions should be part of any lesson plan that involves drawing-based modelling, regardless whether the models are created class-wise (such as in the case described) or individually by the students.

Only informal observations are available from the lessons that were performed with this lesson plan. In all lessons, the class came up with the essential elements for the model after walking through the class. Also in the final discussion, the students were able to indicate the limitations and possible extensions of the model. This shows the value of the final class discussion in which students are stimulated to reflect on the role of the model in understanding the modelled system.

5 Conclusions

The drawing-based modelling approach as introduced in this chapter aims to engage students in genuine scientific activities in which models are created. The approach shows that scientific activities can be challenging and creative. Compared with other approaches to modelling such as those based on coding such as NetLogo [13], or writing equations [11], drawing-based modelling aims at making modelling available for younger children. The development of the SimSketch software is still a work in progress and therefore the results presented here must be considered to be tentative.

Comparing the three cases that were discussed in the previous section, we see that the contexts in which SimSketch was applied vary. The master class students received quite intensive support on their modelling processes, which took place in the context of a really open task. This way of working with students lead to deep thinking about the models that were created and the co-creation of new behaviours in SimSketch. In the new version of SimSketch it will be possible to create new behaviours by writing code, an option that will be available in an "advanced" mode. Adding this option would lead to the possibility of real open-ended modelling, not restricted by the set of available behaviours. In order to make it possible, a library of functions is under construction that would serve as the basis for coding behaviour.

The other two examples are less open ended. Learners were asked to model a specific, given phenomenon. From these examples we can glean that learners are capable of creating models based on a relatively simple instructions and a limited set of behaviours. The collaborative creation of a SimSketch model on traffic is a proof of concept of modelling lessons in which the role of models is introduced and discussed. The main difference between the two situations with respect to the learning goal is the role of the model and phenomenon. In the NEMO study the goal focussed on understanding the domain, the solar system, whereas in the traffic study, modelling itself was in focus. In both cases, drawing-based modelling has a role, but the goals are essentially different. This explains the focus on discussing the role of the model in the traffic study.

On the motivational side, all students indicated they liked working with SimSketch. Whereas this is by no means an indication that they developed a more positive attitude toward science, at least it is a start. The shift in thinking about models by the students in the master class toward a different view on the meaning of models and their function shows that it is possible to reach such a change through working on a modelling task. This makes the master class a proof of concept of the basic ideas of drawing-based modelling as driving force behind the development of scientific attitudes. Of course a single intervention such as the master class or even the single lessons on traffic is not enough to have permanent effect. Science curricula for upper primary and lower secondary grades should include multiple modelling experiences to result in a lasting effect. In order to accomplish this, studies into the conceptualization and actual levels of scientific understanding and attitudes in young students, as well as teacher training programs for school teachers, are necessary [24].

5.1 Future Developments: Creating Scientific Practices on Drawing-Based Modelling

Apart from including more encounters with drawing-based modelling, also improvements in the lesson design are needed, as well as further development of the software. Lesson designs should aim at creating genuine scientific practice within science lessons, with a goal that children see a role for themselves within science. Engaging students this way can turn around the image of science as only deductive and uncreative. Genuine scientific practice includes connection to actual scientific research. In order to achieve this, we are currently developing lessons in collaboration with university researchers. For instance, in collaboration with Naturalis, a museum on natural history, we develop an application for the simulation of evolutionary processes, in which students can create models to explain the evolution of the colours of garden snails. They use data sets made available by a Naturalis researcher to investigate the distribution of the different-coloured snails and then create models simulating the evolution of snails under the selection pressure of birds predating on the snails.

Another development is the introduction of the third dimension in the software. Although drawing is easiest done in two dimensions, some systems to be modelled are intrinsically 3D. For instance in the case of the planetary system, the third dimension is needed to understand that eclipses do not occur every month. Another example is 3D-interaction of molecules based on the spatial distribution of atoms in the molecule. For this, the idea is for students to create the drawing-based model on a 2D canvas and then simulate the model in three dimensions by linking it to a 3D virtual environment. This will allow students to explore more complex systems in a more realistic way.

In conclusion, the results from the present studies show that drawing-based modelling provides a promising approach to engaging young students in science, and to turn around their opinion to see science as a creative enterprise.

References

1. Boerwinkel, D.J., Veugelers, W., Waarlo, A.J.: Burgerschapsvorming, duurzaamheid en natuurwetenschappelijk onderwijs [Citizenship education, sustainability and science education]. Pedagogiek **29**, 155–172 (2009)
2. Hodson, D.: Time for action: science education for an alternative future. Int. J. Sci. Educ. **25** (6), 645–670 (2010). doi:10.1080/09500690305021
3. Osborne, J., Dillon, J.: Science Education in Europe: Critical Reflections, vol. 13. The Nuffield Foundation, London (2008)
4. Bonner, J.F.: Creativity in science. Eng. Sci. **XXII**, 13–17 (1959)
5. Ogborn, J., Wong, D.: A microcomputer dynamical modelling system. Phys. Educ. **19**, 138–142 (1984)
6. Louca, L.T., Zacharia, Z.C.: Modelling-based learning in science education: cognitive, metacognitive, social, material and epistemological contributions. Educ. Rev. **64**, 1–22 (2011). doi:10.1080/00131911.2011.628748
7. Löhner, S., van Joolingen, W., Savelsbergh, E.R., van Hout-Wolters, B.: Student's reasoning during modelling in an inquiry learning environment. Comput. Human Behav. **21**(3), 441–461 (2005). doi:10.1016/j.chb.2004.10.037
8. Quintana, C., Reiser, B.J., Davis, E.A., Krajcik, J., Fretz, E., Duncan, R.G., et al.: A scaffolding design framework for software to support science inquiry. J. Learn. Sci. **13**(3), 337–386 (2004). doi:10.1207/s15327809jls1303_4
9. Windschitl, M., Thompson, J., Braaten, M.: Beyond the scientific method: model-based inquiry as a new paradigm of preference for school science investigations. Sci. Educ. **92**(5), 941–967 (2008). doi:10.1002/sce.20259
10. Giere, R.N.: Science Without Laws. University of Chicago Press, Chicago (1999)
11. Steed, M.: Stella, a simulation construction kit: cognitive process and educational implications. J. Comput. Math. Sci. Teach. **11**(1), 39–52 (1992)
12. Jacobson, M.J., Wilensky, U.: Complex systems in education: scientific and educational importance and implications for the learning sciences. J. Learn. Sci. **15**(1), 11–34 (2006). doi:10.1207/s15327809jls1501_4

13. Wilensky, U., Reisman, K.: Thinking like a wolf, a sheep, or a firefly: learning biology through constructing and testing computational theories—an embodied modelling approach. Cogn. Instruct. **24**(2), 171–209 (2006)
14. Stagecast Software Inc.: Stagecast creator. Retrieved 15 July 2005, from www.stagecast.com (1997)
15. Krajcik, J.S., Sutherland, L.M.: Supporting students in developing literacy in science. Science **328**(5977), 456–459 (2010)
16. Lemke, J.L.: The literacies of science. In: Saul, E.W. (ed.) Crossing Borders in Literacy and Science Instruction: Perspectives on Theory and Practice, pp. 33–47. International Reading Association, Newark (2004)
17. Tversky, B.: Visualizing thought. Topics Cogn. Sci. **3**(3), 499–535 (2011). doi:10.1111/j.1756-8765.2010.01113.x
18. van Joolingen, W., Bollen, L., Leenaars, F., Kenbeek, W.K.: Interactive drawing tools to support modelling of dynamic systems. In: Gomez, K., Lyons, L., Radinsky, J. (eds.) Presented at the 9th International Conference of the Learning Sciences, vol. 2, pp. 184–185. International Society of the Learning Sciences (ISLS), Chicago (2010)
19. Bollen, L., van Joolingen, W.: SimSketch: multi-agent simulations based on learner-created sketches for early science education. IEEE Trans. Learn. Technol. **6**(3), 208–216 (2013). doi:10.1109/TLT.2013.9
20. Mandinach, E.B., Cline, H.F.: Classroom dynamics: the impact of a technology-based curriculum innovation on teaching and learning. J. Educ. Comput. Res. **14**(1), 83–102 (1996)
21. van Joolingen, W., de Jong, T., Lazonder, A.W., Savelsbergh, E.R., Manlove, S.: Co-lab: research and development of an online learning environment for collaborative scientific discovery learning. Comput. Human Behav. **21**(4), 671–688 (2005). doi:10.1016/j.chb.2004.10.039
22. van Joolingen, W., Aukes, A.V.A., Gijlers, H., Bollen, L.: Understanding elementary astronomy by making drawing-based models. J. Sci. Educ. Technol. 1–9 (2014). doi:10.1007/s10956-014-9540-6
23. Vosniadou, S., Skopeliti, I., Ikospentaki, K.: Reconsidering the role of artifacts in reasoning: children's understanding of the globe as a model of the earth. Learn, Instruct. **15**(4), 333–351 (2005). doi:10.1016/j.learninstruc.2005.07.004
24. van Aalderen-Smeets, S., Walma van der Molen, J.: Measuring primary teachers' attitudes toward teaching science: development of the dimensions of attitude toward science (DAS) instrument. Int. J. Sci. Educ. **35**(4), 577–600 (2013). doi:10.1080/09500693.2012.755576

ICT-Enabled Emotional Learning for Special Needs Education

Lennard Chua, Jeremy Goh, Zin Tun Nay, Lihui Huang, Yiyu Cai and Ruby Seah

Abstract Autism Spectrum Disorders (ASD) is a neuro-developmental disability. Children with ASD often find it difficult to express and recognize emotions which make it hard for them to interact socially. Conventional emotional learning for children with ASD often uses medicinal means, and behavioural analysis that could be costly and less effective. There is a significant need to develop Infocomm Technology (ICT) based methods for effective emotional learning. We are interested to develop an iPAD-based game approach for children with ASD to learn emotion through interactive gaming. This chapter reports our project on Social Emotional Learning (SEL) for children with ASDs using an iPAD app.

Keywords Emotional learning · Serious games · Special needs education · Autism spectrum disorders · Facial expression

1 Introduction

Autism Spectrum Disorders (ASD) is a neuro-developmental disability and a life-long disorder that robs many of the affected individuals of independent living, satisfying work and social communication. It can severely stunt their emotional and cognitive development. While the exact cause of ASD is largely unknown, the social gaze disruption theory traces many features of the disorder to a child's inability for processing facial information normally. For instance, children with ASD have difficulty perceiving emotions of other people and processing non-verbal communication. Typically, they appear unaware to gain communication information by attending to others' facial expressions. Often, they exhibit impaired ability to shift attention between objects and faces. These disruptions eventually result in

L. Chua · J. Goh · Z.T. Nay · L. Huang · Y. Cai (✉)
Nanyang Technological University, Singapore, Singapore
e-mail: myycai@ntu.edu.sg

R. Seah
AWWA School, Singapore, Singapore

© Springer Science+Business Media Singapore 2017   29
Y. Cai et al. (eds.), *Simulation and Serious Games for Education*,
Gaming Media and Social Effects, DOI 10.1007/978-981-10-0861-0_3

difficulties socially interacting with others [1, 2]. Early intervention through psychiatric medications and therapies, and behavioural analysis is employed for treatment. These conventional methods are seen not effective in many cases [3] and are usually very expensive [4]. There is a significant need to develop technology-enabled ways for early intervention that can help children with ASD in learning how to express themselves and understand others emotionally.

Game-based learning takes advantages through the use of multi-sensory and engaging approaches. In a controlled game-based learning environment, learners can be guided in their explorations of the unknown by accomplishing learning objectives while entertaining [5]. Due to the portability and touchscreen-enabled interactions, tablet personal computers, particularly iPADs, are now popular [6] and are commonly available in special schools in Singapore to assist learning.

This chapter reports a game-based approach for social emotional learning (SEL). Through the use of actors, mascot and scenes programming, a simple iPAD application is designed for children with ASD to learn emotion. Specifically, the emotions portrayed by local Asian are incorporated in the game for children with ASD to recognize. They can play the game using their iPAD to do emotion learning. Hopefully they can eventually better manage their feelings, and also communicate better with others which are important parts of SEL.

2 Literature Review

Social deficits are a major feature of ASD according to American Psychiatric Association [7]. There is evidence suggesting that individuals with ASD have difficulty recognizing facial expressions of emotion [8, 9]. Many studies indicate that under the broad category of socialization deficits, individuals with ASD struggle with emotions. Back et al. [10], Baron-Cohen et al. [11] and Wallace et al. [12] found that individuals with ASD performed significantly worse than their typically developing peers in: (a) determining affect information as shown in the eyes versus the whole face; (b) labelling complex emotions when displayed through the eyes only; and (c) recognizing the emotion 'fear' from the eyes and 'disgust' from the mouth when presented with partial faces, respectively. On the contrary, Boraston et al. [13] and Humphreys et al. [14] reported that individuals with ASD scored no difference with the control group when the emotions being depicted were 'scared' and 'disgust,' but they did significantly worse on all other emotions; participants with ASD did as well as controls on recognizing emotions, except for the emotion of 'fear' which was often confused with 'surprise' [15]. According to Baron-Cohen et al. [16] and Kaland et al. [17] in the Theory of Mind (ToM) high-functioning autism (HFA) and Asperger syndrome (AS) are characterized by impairment in social interaction, such as difficulty in developing friendships and a lack of understanding of emotions and minds of others. Klucharev and Sams [18] highlighted that understanding emotions usually requires multi-sensory processing. Emotions and the feelings of others are interpreted from speech prosody as well as from body and facial gestures. The ability to recognize other people's emotional state requires the ability to divide attention and focus gaze on relevant information. This processing is mostly subconscious [19]. A deficit in understanding private events has been linked with impairments in social behaviours [20]. Durand [21], however, found that the inability to label private events was linked to inappropriate behaviours. Thus, developing training procedures to remediate these difficulties is a clear priority. Harris et al. [22] showed children with ASD have a limited ability to label emotions provoked by particular situations. Several studies have found that children with ASD were less able than their age-matched, typically developing counterparts to recognize emotional reactions in certain situations [23]. To increase the likelihood that each child would learn a generalized repertoire of emotion understanding, multiple exemplars of emotion identification were trained using a multi-component procedure [24]. Self-regulation [25] is also a problem for many children with special needs. Children with developmental conditions, such as attention-deficit hyperactivity disorder (ADHD), or physical conditions, such as cerebral palsy, have limited power to will their behaviours and consequent self-appraisals may be negative or overtly defensive. Motivation [26, 27] is important for all children, especially for children with special needs. In nearly all cultures, there are six commonly recognized basic expressions of facial affect: happiness, sadness, anger, fear, surprise and disgust [28, 29] which typically developing children can accurately identify and label by age 3 [30]. Children with ASD [31–36] are less accurate compared to the typically developing children in their ability to identify these basic facial expressions of emotion from photographs, cartoons and video clips. Involuntary facial expression processing has been investigated by Dahl [37] in which participants were required to categorize words as positive or negative, while ignoring the positive or negative expression upon which these words were superimposed. Beall and Herbert [38] recently showed larger interference effects for word targets with face distractors than word distractors with face targets, concluding that emotional faces are processed more 'automatically' than emotional words [39]. In reality, facial expressions are rarely displayed at their maximum intensity. Few studies took emotion intensity into account focusing on emotion-processing abnormalities (i.e. bias or insensitivity) in clinical populations such as autism adults [40], children with psychopathic tendencies [41] and with social phobia [42] or schizophrenia [43]. Except Philippot and Douilliez [40], these studies suggested that certain emotions (i.e. fear) may require a higher intensity of expression before they are correctly recognized by the clinical group examined. An understanding of the normal development of the ability to recognize more subtle displays (i.e. lower intensity) of facial expression remains extremely limited, yet may prove valuable in detecting early aberrant patterns in emotion processing [44]. The ability to quickly assess one's social environment, taking into account of another's identity and emotional cues, is crucial for successful social communication and development [45]. Developmental studies of emotion processing indicated that preschool children's explicit recognition of emotional expressions emerges over development, with happiness recognized earliest and most accurately, followed by sad or angry expressions, then by surprise or fear [46].

Traditional teaching for children with ASD and their difficulties was discussed by Sharmin et al. [44], Baskett [47], Kanner [48] and Rahman et al. [49]. Kerr et al. [50] and Tartaro [51] explored the effect of technological environments on the process of assisting children with autism. Lim et al. [52] discussed role-play for education. Games allow players to escape into fantasy worlds, encourage exploration of excitements of things, people and places that are otherwise inaccessible in the real world, inducing a 'suspension of disbelief' in players. Learning often takes place during the game play, with immediate feedback. Crawford [53] found that the subject to be learned is directly related to the game environment where constant cycles of hypothesis formulation, testing and revision are evoked as the gamer experiences continuous cycles of cognitive disequilibrium and resolution. Shukla-Mehta et al. [54] showed that children with ASD learn better through interactive visual means. Ramdoss et al. [55] provided a systematic analysis for the use of Computer-based Intervention (CBI) in enhancing communication skills in autistic children. Recently, there was significant effort in the direction of using modern technology to develop computer-based systems which can be used to teach children with ASD various social and communication skills. Tanaka et al. [56] designed a program known as 'Let's Face It'—which uses computerized games in order to teach basic face processing skills to children suffering from ASD. One component of their system used facial expressions of the children as an input to the game which involves solving a maze [57]. Kerstin [58] and Jain et al. [59] described their work using the robot and several facial expressions (smile, frown, laugh, etc.) for children with ASD. Virtual environments, interactive technology and virtual peers are discussed by Rahman et al. [60], Sharmin et al. [44], Anwar et al. [61], Alqahtani et al. [62], Cai et al. [63], Simpson [64], and Mary and Thormann [65] for children with autism. Lehman [66], Michaud and Théberge-Turmel [67], Pars et al. [68], Cai and Goei [69] and Kerr et al. [50] discussed learning through playing or serious gaming in special needs education. Play incorporates some of the most important behaviours for a full cognitive development of children in different areas [70, 71]. Children with cognitive and social disabilities are often unable to engage in play activities with their peers, because it is difficult for them to establish relationships and to explore their social environment [72]. They prefer to play alone or in limited dyadic interactions. In everyday social life the ability to engage in activities requiring shared attention is the key to understand the interaction partner as an intentional agent and to form cooperative strategies based on this understanding. Spitzer [73] and Lehmann et al. [74] looked into the function of play for children with autism, to develop and improve this ability. Virtual Reality (VR) is an emerging technology for learning and teaching including special needs education [75, 76, 77]. In the form of simulated environment it helps the children to create a coherent organization of certain important everyday activities [78]. Dautenhahn [79] hoped that such virtual learning environments will aid the rehabilitation process, and through refinement for each individual patient provide an enjoyable 'toy' and increase quality of life. Moore et al. [80] used a collaborative virtual environment (CVE) to analyze the ability of children with autism to understand basic emotions represented by a humanoid avatar. They carried out two empirical studies, whose results suggest that children with autism maybe able to accurately recognize emotions in virtual representations.

3   iPAD App Design for Emotional Learning

Children with ASD often lack the ability to relate to people in usual ways. They encounter difficulty communicating with people using non-verbal cues leading to failure in building peer relationships partially due to their lack of ability to recognize emotions. They are incapable of expressing or reciprocating to people socially being not able to understand or predict the emotions of people around. Their ability in recognizing and showing their feelings through facial expressions is also very limited. As Joseph and Tanaka [81] described, in processing information about the human face, they may overemphasize one part of the face such as the mouth rather than attending to its overall shape or focusing on the eyes as most children do. They tend also to avoid eye-to-eye gaze and not play with others.

Children's emotional states interweave with the cognitive dynamics of the learning process. The emotional state is linked to children's learning outcome. When children with ASD have positive emotion, they are often happy, satisfied and even curious in their learning that can help them become active learners with hopeful and interesting ideas to explore. Negative emotion, however, can lead children with ASD disappointed, confused and disturbed in their learning frustrated with discarding and misconceptions.

Emotional learning therefore is particularly important to children with ASD. Motivating children with ASD to learn positively and actively can accumulate skills to achieve a task in the next levels of the learning [82]. This incites and encourages them to learn the necessary elements of education. In the work of Malone and Lepper [83] on the characteristic motivating factors of games, recognition is highlighted as a source of motivation. More specifically, it involves making the results of activities performed by players visible to others.

CAT Kit developed by Tony Attwood is widely used as a training kit [84] in special needs education. Figure 1 shows the CAT Kit used in special needs education to 'check in' the mood of children with ASD. With the kit, children will be invited to choose the description of the emotion that they are feeling, accompanied by a sticker of how their face would look like when experiencing that particular emotion. The range of emotions available in the CAT Kit includes joy, sorrow, safety, fear, affection, anger, pride, shame and surprise.

There are several shortcomings on using this CAT Kit. First, children with ASD may have difficulty to relate to the emotions portrayed by the drawings with the kit as they are not life-like enough. Second, the kit is designed for manual work therefore with limited function of interaction. Third, the kit (USD 199.95) can only be accessed by one child at a time. This motivates us to design an iPAD application with features such as life-like, easy duplicate, and, more importantly, good

Fig. 1 Mood check in for
CAT kit

interaction to assist children with ASD for better emotional learning. iPADs (Fig. 2) are useful for children with disabilities because some of the applications available for them replace bulky, more expensive forms of assistive technology [6].

The structure of the iPAD game is designed such that it will provide a flow to all the different levels of emotional learning. Each world will have six levels for players to complete before proceeding to the next. Likewise for the levels, players are only allowed to proceed when they get the correct answer. Figure 3 is a schematic diagram of the structure of the iPAD app. World selection is needed for easy navigation by players as the number of levels increases. A different theme will be designed for each world to enhance their visual acceptance. The level design and the layout design will be simple and concise as children with ASD may have a short attention span. When the game starts, players will be greeted by a friendly mascot/avatar (Fig. 4).

Role-play is an ideal tool for helping children with ASD to develop and practice communication and social skills. The game provides a safe and structured alternative of role-plays using videos of different situations to express the different kinds of emotions involved. For example, a child fell down and got hurt (Fig. 5) is experiencing pain and sadness. These videos are staged and can be replayed without injuring any parties and actors involved. It can also act as a teaching tool for

Fig. 2 iPAD for ASD children

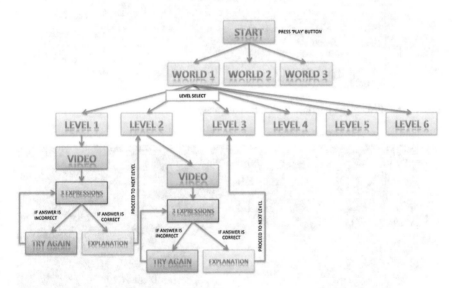

Fig. 3 Schematic diagram of game structure

teachers, caregivers, parents or peers to re-enact scenarios that the child has experienced before and be relate to the emotions embedded. Currently, as most of the existing games and teaching tools for children with ASD are usually non-Asian, this makes it not straightforward for local children to relate their emotions to unfamiliar faces. Using an Asian actor in the videos and the same actor in the pictures will provide better perspectives to how the actor is feeling.

Colour is a powerful design element that produces profound psychological and physiological reactions. Some children with ASD may be more sensitive to colour in the learning environment due to heightened sensory responses and strong visual

Fig. 4 The iPAD game interface with a friendly mascot

Fig. 5 An Asian girl acting a scenario

Fig. 6 I feel SAD

processing abilities [85]. Therefore colours used in this game should be able to relate to the particular emotion and seek the user's attention and prolonging his or her attention span. For instance, the game designed uses 'grey' to portray the negative emotions involved [86], and 'green' regarded as the most restful and pleasing for the eye [87]. Figure 6 is a snapshot of the game on "I feel SAD".

4 Experiment

The purpose of conducting this experiment was to see the effectiveness of the game play of the iPAD app as a tool for emotional learning of children with ASD. The experiment carried out actual test on children with ASD to see their reaction and behaviour before, during and after playing the game. Their reaction and behaviour would be crucial to improve the game design and development.

The experimental was conducted in AWWA School at the ICT room. Six children with ASD participated in the experiment. The experiment would be assessed by one psychologist and two class teachers from AWWA. Each experiment session had one child at a time and the child was first allowed to work his/her way to the iPAD app (free interaction game play). This is to see if the game was intuitive and allowed the child to play without any external help (Fig. 7a). Next,

(a)	(b)

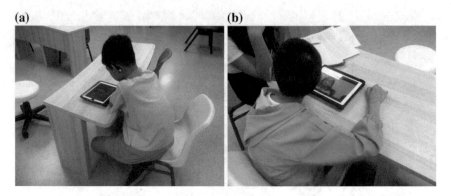

Fig. 7 Children with ASD playing the iPAD game: **a** without any help; **b** with assistance

PARTICIPANT ASSESSMENT CHART

No	Objectives	Rating Scale					Problem faced?	Other remarks
Student:		Date						
Assessor:		Assessor's Appointment:						
		POOR		FAIR		EXCELLENT		
1	Attention – app able to grab attention of the user	1	2	3	4	5		
2	Intuitiveness – user able to perform the task on their own without supervision	1	2	3	4	5		
3	Attention – user able to maintain focus with task for a period of time	<6mins	6mins	8mins	10mins	>10mins		
4	Effort – user able to make an effort to answer correctly	1	2	3	4	5		
5	Comparing – user able to differentiate the expressions	1	2	3	4	5		
6	Language – user able to identify the expressions	1	2	3	4	5		
7	Tolerance – user able to wait for in between levels	1	2	3	4	5		
8	Self-Esteem – user able to display confidence in successful attempts	1	2	3	4	5		
9	Self-Esteem – user able to continue with confidence despite failures/mistakes	1	2	3	4	5		
10	Interest - user shows signs of anticipation of more levels	1	2	3	4	5		
Q	Any improvements that can be made?							

Fig. 8 Assessment criteria and scoring

help was gradually induced (Fig. 7b) to make sure that the child stays within the app and at the same time, teaching the emotions embedded within the videos. Major emphasis was placed on the attention span on the app. Throughout the experiment, the ICT teacher, along with a psychologist, assessed the participating children with ASD according to the assessment sheet provided in Fig. 8, while a class teacher assisted them in handling the child.

The following criteria were formulated to assess the effectiveness of the game:

- Attention: whether the game was able to grab attention of the player
- Intuitiveness: whether the player was able to perform tasks on their own without supervision
- Effort: whether the player was making effort to answer
- Body Language: whether the player was able to identify the expressions
- Interest: whether the player showed signs of anticipation
- Tolerance: whether the player was able to wait in between levels
- Comparison: whether the player was able to differentiate expressions

The outcome of the experiment was of mixed results, as the group of participants with ASD is of different autistic levels. Some were of high function ASD and a few of intermediate ASD level, hence, those with high function ASD were able to

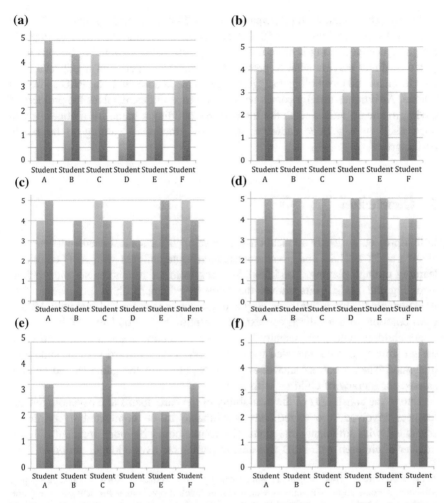

Fig. 9 Assessment results for two tests: *a* attention, *b* intuitiveness, *c* effort, *d* tolerance, *e* comparison, and *f* self-esteem

complete the experiment without much trouble but those with intermediate level ASD had some difficulty with the experiment. The troubles encountered by the intermediate level ASD, ranging from the inability to concentrate for long period of time to the inability to follow proper instructions. However, given proper guidance by the psychologist and the attention they needed, they were able to complete the experiment. They were also more willing to follow instructions given by the psychologist and they were able to express their feelings about the game when asked by the psychologist. Figure 9 shows the assessment results of two tests on Attention, Intuitiveness, Effort, Tolerance, Comparison and Self-esteem.

In this experiment, each participant only has 6–10 mins to play and results were gathered from these observations. The experimental results were used to improve the game design making it more user friendly. For instance, blinking buttons were added to direct players what to press. Also from this experiment, feedbacks from both teachers and psychologist suggested the game to come in both iPAD as well as iPhone. As iPhones usually comes with the 4G capabilities, it solves the problem of having to stream videos from the Internet. By building a game for iPhones, it will open itself to a greater audience and allow emotional learning to be taught on-the-go.

5 Conclusion

In this project, an iPAD application was developed for special needs education. It aims to assist teachers and psychologists to educate children with ASD emotional learning through playing. The initial feedback from the psychologists and teachers assessing the participants shows the game-based approach is promising. This iPAD-enabled game provides an easy and interactive way for children with ASD to learn emotion with localized context. The solution has a good scalability.

As children with ASD usually have a short attention span, a mini style game design strategy is recommended for future work. Better reward systems should be integrated when a correct answer is given. Narration of the game should be considered for literacy of children with ASD.

Due to the constraints of the availability of the participants and psychologist and teachers involved, the experiment was unable to be conducted as originally scheduled with sufficient time slots. The results could be more apparent if larger number of children with ASD was able to participate over a longer period of time.

Acknowledgments This project is supported by The Temasek Trust Funded Singapore Millennium Foundation. The authors would like to thank the volunteer for acting in the game. Thanks go to Mr. Izad, Ms. Christine Kuek, Mdm. Gurmit, Professor Sui Lin Goei and many other people who help in this project in one way or another.

References

1. Takacs, B.: Special education and rehabilitation: teaching and healing with interactive graphics. IEEE Comput. Graph. Appl. **25**(5), 40–48 (2005). doi:10.1109/mcg.2005.113
2. Ramachandran, V.S., Oberman, L.M.: Broken mirrors: a theory of autism. Sci. Am. **295**(5), 62–69 (2006)
3. Ospina, M.B., Seida, J.K., Clark, B., Karkhaneh, M., Hartling, L., Tjosvold, L., Smith, V.: Behavioural and developmental interventions for autism spectrum disorder: a clinical systematic review. PLoS ONE **3**(11), e3755 (2008). doi:10.1371/journal.pone.0003755
4. Sharpe, D.L., Baker, D.L.: Financial issues associated with having a child with autism. J. Fam. Econ. Issues **28**(2), 247–264 (2007). doi:10.1007/s10834-007-9059-6

5. Muñoz, K., Kevitt, PMc, Lunney, T., Noguez, J., Neri, L.: An emotional student model for game-play adaptation. Entertainment Comput. **2**(2), 133–141 (2011). doi:10.1016/j.entcom.2010.12.006
6. Shah, N.: Special education pupils find learning tool in iPad applications. Educ. Week **30**(22), 1 (2011)
7. American Psychiatric Association: Diagnostic and statistical manual of mental disorders, 4, DSM-IV-TR edn. American Psychiatric Association, Washington, DC (2000)
8. Harms, M.B., Martin, A., Wallace, G.L.: Facial emotion recognition in autism spectrum disorders: a review of behavioral and neuroimaging studies. Neuropsychol. Rev. **20**(3), 290–322 (2010). doi:10.1007/s11065-010-9138-6
9. Spencer, M.D., Holt, R.J., Chura, L.R., Suckling, J., Calder, A.J., Bullmore, E.T., Baron-Cohen, S.: A novel functional brain imaging endophenotype of autism: the neural response to facial expression of emotion. Transl Psychiatry **1**, e19 (2011). doi:10.1038/tp.2011.18
10. Back, E., Ropar, D., Mitchell, P.: Do the eyes have it? Inferring mental states from animated faces in autism. Child Dev. **78**(2), 397–411 (2007). doi:10.1111/j.1467-8624.2007.01005.x
11. Baron-Cohen, S., Wheelwright, S., Jolliffe, T.: Is there a "Language of the Eyes"? Evidence from normal adults, and adults with autism or Asperger syndrome. Vis Cogn. **4**(3), 311–331 (1997). doi:10.1080/713756761
12. Wallace, S., Coleman, M., Bailey, A.: An investigation of basic facial expression recognition in autism spectrum disorders. Cogn. Emot. **22**(7), 1353–1380 (2008). doi:10.1080/02699930701782153
13. Boraston, Z., Blakemore, S.J., Chilvers, R., Skuse, D.: Impaired sadness recognition is linked to social interaction deficit in autism. Neuropsychologia **45**(7), 1501–1510 (2007). doi:10.1016/j.neuropsychologia.2006.11.010
14. Humphreys, K., Minshew, N., Leonard, G.L., Behrmann, M.: A fine-grained analysis of facial expression processing in high-functioning adults with autism. Neuropsychologia **45**(4), 685–695 (2007). doi:10.1016/j.neuropsychologia.2006.08.003
15. Caldeira, M., Edmunds, A.: Inconsistencies in autism-specific emotion interventions: cause for concern. Exceptionality Educ. Int. **22**(1), 17–36 (2012)
16. Baron-Cohen, S., Leslie, A.M., Frith, U.: Does the autistic child have a "theory of mind"? Cognition **21**(1), 37–46 (1985)
17. Kaland, N., Smith, L., Mortensen, E.L.: Response times of children and adolescents with Asperger syndrome on an 'advanced' test of theory of mind. J. Autism Dev. Disord. **37**(2), 197–209 (2007). doi:10.1007/s10803-006-0152-8
18. Klucharev, V., Sams, M.: Interaction of gaze direction and facial expressions processing: ERP study. NeuroReport **15**(4), 621–625 (2004)
19. Kuusikko, S., Haapsamo, H., Jansson-Verkasalo, E., Hurtig, T., Mattila, M.-L., Ebeling, H., Moilanen, I.: Emotion recognition in children and adolescents with autism spectrum disorders. J. Autism Dev. Disord. **39**(6), 938–945 (2009). doi:10.1007/s10803-009-0700-0
20. LeBlanc, L.A., Coates, A.M., Daneshvar, S., Charlop-Christy, M.H., Morris, C., Lancaster, B.M.: Using video modeling and reinforcement to teach perspective-taking skills to children with autism. J. Appl. Behav. Anal. **36**(2), 253–257 (2003)
21. Durand, V.: Functional communication training using assistive devices: effects on challenging behavior and affect. Augmentative Altern. Commun. **9**(3), 168–176 (1993). doi:10.1080/07434619312331276571
22. Harris, P.L., Johnson, C.N., Hutton, D., Andrews, G., Cooke, T.: Young children's theory of mind and emotion. Cogn. Emot. **3**(4), 379–400 (1989). doi:10.1080/02699938908412713
23. Downs, A., Smith, T.: Emotional understanding, cooperation, and social behavior in high-functioning children with autism. J. Autism Dev. Disord. **34**(6), 625–635 (2004)
24. McHugh, L., Bobarnac, A., Reed, P.: Brief report: teaching situation-based emotions to children with autistic spectrum disorder. J. Autism Dev. Disord. **41**(10), 1423–1428 (2011). doi:10.1007/s10803-010-1152-2

25. Malmivuori, M.-L.: Affect and self-regulation. Educ. Stud. Math. **63**(2), 149–164 (2006). doi:10.1007/s10649-006-9022-8
26. Hannula, M.S.: Motivation in mathematics: goals reflected in emotions. Educ. Stud. Math. **63**(2), 165–178 (2006). doi:10.1007/s10649-005-9019-8
27. Rodd, M.: Commentary: mathematics, emotion and special needs. Educ. Stud. Math. **63**(2), 227–234 (2006). doi:10.1007/s10649-005-9014-0
28. Ekman, P., Friesen, W.V.: Measuring facial movement. Environ. Psychol Nonverbal Behav. **1**(1), 56–75 (1976). doi:10.1007/BF01115465
29. Ekman, P.: Basic emotions. In: Handbook of Cognition and Emotion, pp. 45–60. Wiley, New Jersey (2005)
30. Widen, S.C., Russell, J.A.: A closer look at preschoolers' freely produced labels for facial expressions. Dev. Psychol. **39**(1), 114–128 (2003)
31. Bal, E., Harden, E., Lamb, D., Van Hecke, A.V., Denver, J.W., Porges, S.W.: Emotion recognition in children with autism spectrum disorders: relations to eye gaze and autonomic state. J. Autism Dev. Disord. **40**(3), 358–370 (2010). doi:10.1007/s10803-009-0884-3
32. Wright, B., Clarke, N., Jordan, J.O., Young, A.W., Clarke, P., Miles, J., Williams, C.: Emotion recognition in faces and the use of visual context Vo in young people with high-functioning autism spectrum disorders. Autism **12**(6), 607–626 (2008). doi:10.1177/1362361308097118
33. Celani, G., Battacchi, M.W., Arcidiacono, L.: The understanding of the emotional meaning of facial expressions in people with autism. J. Autism Dev. Disord. **29**(1), 57–66 (1999)
34. Gross, T.F.: The perception of four basic emotions in human and nonhuman faces by children with autism and other developmental disabilities. J. Abnorm. Child Psychol. **32**(5), 469–480 (2004)
35. Tardif, C., Laine, F., Rodriguez, M., Gepner, B.: Slowing down presentation of facial movements and vocal sounds enhances facial expression recognition and induces facial-vocal imitation in children with autism. J. Autism Dev. Disord. **37**(8), 1469–1484 (2007). doi:10.1007/s10803-006-0223-x
36. Williams, B.T., Gray, K.M., Tonge, B.J.: Teaching emotion recognition skills to young children with autism: a randomised controlled trial of an emotion training programme. J. Child Psychol. Psychiatry **53**(12), 1268–1276 (2012). doi:10.1111/j.1469-7610.2012.02593.x
37. Dahl, M.: Asymmetries in the processing of emotionally valenced words. Scand. J. Psychol. **42**(2), 97–104 (2001). doi:10.1111/1467-9450.00218
38. Beall, P.M., Herbert, A.M.: The face wins: stronger automatic processing of affect in facial expressions than words in a modified Stroop task. Cogn. Emot. **22**(8), 1613–1642 (2008). doi:10.1080/02699930801940370
39. Baggott, S., Palermo, R., Fox, A.M.: Processing emotional category congruency between emotional facial expressions and emotional words. Cogn. Emot. **25**(2), 369–379 (2011). doi:10.1080/02699931.2010.488945
40. Philippot, P., Douilliez, C.: Social phobics do not misinterpret facial expression of emotion. Behav. Res. Ther. **43**(5), 639–652 (2005). doi:10.1016/j.brat.2004.05.005
41. Blair, R.J.R., Colledge, E., Murray, L., Mitchell, D.G.V.: A selective impairment in the processing of sad and fearful expressions in children with psychopathic tendencies. J. Abnorm. Child Psychol. **29**(6), 491–498 (2001). doi:10.1023/A:1012225108281
42. Kohler, C.G., Turner, T.H., Bilker, W.B., Brensinger, C.M., Siegel, S.J., Kanes, S.J., Gur, R.C.: Facial emotion recognition in schizophrenia: intensity effects and error pattern. Am. J. Psychiatry **160**(10), 1768–1774 (2003)
43. Herba, C.M., Landau, S., Russell, T., Ecker, C., Phillips, M.L.: The development of emotion-processing in children: effects of age, emotion, and intensity. J. Child Psychol. Psychiatry **47**(11), 1098–1106 (2006). doi:10.1111/j.1469-7610.2006.01652.x
44. Sharmin, M.A., Rahman, M.M., Ahmed, S.I., Rahman, M.M., Ferdous, S.M.: Teaching intelligible speech to the autistic children by interactive computer games. Paper presented at the proceedings of the 2011 ACM symposium on applied computing, TaiChung, Taiwan (2011)

45. Cunningham, J.G.: Differential salience of facial features in children's perception of affective expression. Child Dev. **57**, 136–142 (1986)
46. Camras, L.A., Allison, K.: Children's understanding of emotional facial expressions and verbal labels. J. Nonverbal Behav. **9**(2), 84–94 (1985). doi:10.1007/BF00987140
47. Baskett, C.B.: The Effect of Live Interactive Video on the Communicative Behavior in Children with Autism. University of North Carolina, Chapel Hill (1996)
48. Kanner, L.: Autistic disturbances of affective contact. Acta Paedopsychiatrica **35**(4), 100–136 (1968)
49. Rahman, M., Ferdous, S.M., Ahmed, S.I., Anwar, A.: Speech development of autistic children by interactive computer games. Interact. Technol. Smart Educ. **8**(4), 208–223 (2011)
50. Kerr, S.J., Neale, H.R., Cobb, S.V.G.: Virtual environments for social skills training: the importance of scaffolding in practice. Paper presented at the Proceedings of the fifth international ACM conference on Assistive technologies, Edinburgh, Scotland (2002)
51. Tartaro, A.: Storytelling with a virtual peer as an intervention for children with autism. SIGACCESS Accessibility Comput. **84**, 42–44 (2006). doi:10.1145/1127564.1127573
52. Lim, M.Y., Kriegel, M., Aylett, R., Enz, S., Vannini, N., Hall, L., Leichtenstern, K.: Technology-enhanced role-play for intercultural learning contexts. In: Natkin, S., Dupire, J. (eds.) Entertainment Computing—ICEC 2009, vol. 5709, pp. 73–84. Springer, Berlin, Heidelberg (2009)
53. Crawford, C.: The art of computer game design, p. c1984. Osborne/McGraw-Hill, Berkeley (1984)
54. Shukla-Mehta, S., Miller, T., Callahan, K.J.: Evaluating the effectiveness of video instruction on social and communication skills training for children with autism spectrum disorders: a review of the literature. Focus Autism Other Dev. Disabil. **25**(1), 23–36 (2010)
55. Ramdoss, S., Lang, R., Mulloy, A., Franco, J., O'Reilly, M., Didden, R., Lancioni, G.: Use of computer-based interventions to teach communication skills to children with autism spectrum disorders: a systematic review. J. Behav. Educ. **20**, 55–76 (2011)
56. Tanaka, J.W., Wolf, J.M., Klaiman, C., Koenig, K., Cockburn, J., Herlihy, L., Brown, C., Stahl, S., Kaiser, M.D., Schultz, R.T.: Using computerized games to teach face recognition skills to children with autism spectrum disorder: the let's face it! Program. J. Child Psychol. Psychiatry **51**, 944–952 (2010)
57. Cockburn, J., Bartlett, J., Tanaka J.W., Movellan, J., Pierce, J., Schultz, R.: Smilemaze: a tutoring system in real-time facial expression perception and production for children with autism spectrum disorder. In: International Conference on Automatic Face and Gesture Recognition 2008
58. Dautenhahn, K., Nehaniv, C.L., Walters, M.L., Robins, B., Kose-Bagci, H., Mirza, N.A., Blow, M.: Kaspar—a minimally expressive humanoid robot for human-robot interaction research. Appl. Bionics Biomech. **6**, 369–397 (2009)
59. Jain, S., Tamersoy, B., Zhang, Y., Aggarwal, J.K., Orvalho, V.: An interactive game for teaching facial expressions to children with autism spectrum disorders. Paper presented at the 2012 5th international symposium on communications control and signal processing (ISCCSP) 2–4 May 2012
60. Rahman, M.M., Ferdous, S.M., Ahmed, S.I.: Increasing intelligibility in the speech of the autistic children by an interactive computer game. Paper presented at the 2010 IEEE international symposium on multimedia (ISM) 13–15 Dec 2010
61. Anwar, A., Rahman, M.M., Ferdous, S.M., Anik, S.A., Ahmed, S.I.: A computer game based approach for increasing fluency in the speech of the autistic children. Paper presented at the 2011 11th IEEE international conference on advanced learning technologies (ICALT) 6–8 July 2011

62. Alqahtani, A., Jaafar, N., Alfadda, N.: Interactive speech based games for autistic children with Asperger syndrome. Paper presented at the 2011 international conference and workshop on current trends in information technology (CTIT) 26–27 Oct 2011
63. Cai, Y., et al.: Design and development of a virtual dolphinarium for children with autism. IEEE Trans. Neural Syst Rehabil. Eng. Publ. IEEE Eng. 21(2), 208–217 (2013)
64. Simpson, E.S.: Video games as learning environments for students with learning disabilities. Children Youth Environ. 19(1), 306–319 (2009)
65. Seegers, M., Thormann, J.: Special technological possibilities for students with special needs. Learn. Lead. Technol. 29(3) (2001)
66. Lehman, J.F.: Toward the use of speech and natural language technology in intervention for a language-disordered population. Paper presented at the Proceedings of the third international ACM conference on Assistive technologies, Marina del Rey, California, USA (1998)
67. Michaud, F., Théberge-Turmel, C.: Mobile robotic toys and autism. In: Socially Intelligent Agents, p. 125 (2002)
68. Pars, N., Carreras, A., Durany, J., Ferrer, J., Freixa, P., Gmez, D., Sanjurjo, L.: Promotion of creative activity in children with severe autism through visuals in an interactive multisensory environment. Paper presented at the IDC '05: proceeding of the 2005 conference on interaction design and children, Boulder, Colorado (2005)
69. Cai, Y., Goei, S.L. (eds.): Simulation and Serious Games. Springer, Berlin Dec 2013
70. World Health Organization: International Classification of Functioning, Disability and Health. World Health Organization, Geneva (2001)
71. Power, T.G.: Play and Exploration in Children and Animals. Lawrence Erlbaum Associates, Mahwah (2000)
72. Besio, S. (ed.): Analysis of Critical Factors Involved in Using in Interactive Robots for Education and Therapy of Children with Disabilities. UNIService, Trento (2008)
73. Spitzer, S.L.: Play in children with autism: structure and experience. In: Parham, L.D., Fazio, L.S. (eds.) Play in Occupational Therapy for Children, 2nd edn, pp. 351–374. Mosby Elsevier, St. Louis (2008)
74. Lehmann, H., Iacono, I., Robins, B., Marti, P., Dautenhahn, K.: Make it move: playing cause and effect games with a robot companion for children with cognitive disabilities. Paper presented at the proceedings of the 29th annual european conference on cognitive ergonomics, Rostock, Germany (2011)
75. Veltjen, A.: Using games to support students with special needs! In: Proceedings of the European Conference on Games Based Learning, p. 512 (2010)
76. Cai, Y.: When VR Meets Special Needs Education. 2014 International Forum on Assistive Technology for Learning, Xian, China, 21–22 Nov 2014
77. Cai, Y.: The wonders of virtual reality technology as a teaching tool for early childhood teachers: an interactive demonstration. In: YWCA Inaugural Conference on Person-Centredness and Inclusive Early Childhood Education, Singapore, 18 Sept 2015
78. Charitos, D., Karadanos, G., Sereti, E., Triantafillou, S., Koukouvinou, S., Martakos, D. Employing virtual reality for aiding the organisation of autistic children behaviour in everyday tasks (2000)
79. Dautenhahn, K.: Design issues on interactive environments for children with autism. Paper presented at the proceedings of ICDVRAT 2000, the 3rd international conference on disability, virtual reality and associated technologies (2000)
80. Moore, D., Cheng, Y., McGrath, P., Powell, N.: Avatars and autism, assistive technologies. In: Pruski, A., Knops, H. (eds.) Virtuality to Reality, pp. 442–448. IOS Press (2005)
81. Joseph, R.M., Tanaka, J.: Holistic and part based face recognition in children with autism. J. Child Psychol. Psychiatry 44, 529–542 (2003)
82. Prensky, M.: Digital Game-Based Learning. McGraw-Hill, New York (2001)
83. Malone, T.W., Lepper, M.R.: Making learning fun: a taxonomy of intrinsic motivations for learning. In: Snow, R.E., Farr, M.J. (eds.) Erlbaum, vol. 3, pp. 223–253. (1987)

84. Cognitive Affective Training Kit. http://www.catkit-us.com/
85. Freed, J., Parsons, L.: Right-Brained Children in a Left-Brained World. Fireside, New York (1997)
86. Boyatzis, C.J., Varghese, R.: Children's emotional associates with colors. J. Genetic Psychol. **155**(1), 77–85 (1993)
87. Karp, E.M., Karp, H.B.: Color associations of male and female fourth-grade school children. J. Psychol. **122**(4), 383–388 (2001)

Simulation-Enabled Vocational Training for Heavy Crane Operations

Panpan Cai, Indhumathi Chandrasekaran, Yiyu Cai, Yong Chen
and Xiaoqun Wu

Abstract A simulation system for vocational training of crane operation is presented in this chapter aiming to improve the lifting personnel's operational skills and safety awareness. The system utilizes various types of ICT such as CAD modeling, Smart Plants, and Point Cloud. The stereographic rendering technique and multiple types of interactions supported by the system improve the level of realism for crane operation training. Interactive operation designed with the simulation system helps trainees to gain operation skills which hopefully will be translated to productivity in their work.

Keywords Vocational training · Crane · Lifting · Simulation

1 Introduction

Vocational training is important for industries like transportation, aviation, construction, health care, and so on. These industries require skilled workers to conduct highly experienced and knowledge-based operations. Vocational training, which is usually conducted by vocational training centers, technical schools, or community colleges, develops expertise in the industries through a combination of school-based and workplace training. In particular, the lifting business requires a large number of skilled workers to perform crane driving, crane lifting, rigging, and other tasks. The complex and hazardous nature of these operations makes the lifting training an intensive and critical process. Professional and continuing training can help the lifting workforce keep upgrading of their knowledge and skills thus improving safety and productivity.

P. Cai · I. Chandrasekaran · Y. Cai (✉) · Y. Chen · X. Wu
School of Mechanical and Aerospace Engineering, Nanyang Technological University,
Singapore, Singapore
e-mail: myycai@ntu.edu.sg

Y. Cai · Y. Chen
Institute of Media Innovation, Nanyang Technological University, Singapore, Singapore

© Springer Science+Business Media Singapore 2017 47
Y. Cai et al. (eds.), *Simulation and Serious Games for Education*,
Gaming Media and Social Effects, DOI 10.1007/978-981-10-0861-0_4

1.1 Heavy Lifting

Heavy lifting operations are common in petrochemical, pharmaceutical, and construction industries. Heavy cranes with capacities from a few hundreds to thousand tons are used to lift boilers, heaters, construction materials, and so on in industry sites. Heavy lifting often has high safety risks because of the large weight of the loads involved, complexity of the environments, and complicated operation procedures. Thus, it requires a big group of professional managers and workers cooperating with each other to finish a heavy lifting job.

A typical heavy lifting project contains three phases: investigation, planning, and execution. At the very beginning of the project, a group of people investigate the work site or plant by analyzing its environment and ground conditions. For old plants who do not have digital versions, the group will use estimated CAD drawings referring pictures taken from the site. Other plants may have digital models in CAD, Smart Plants, or PDMS (Plant Design Management System) formats. These digital data would be much more detailed and accurate which can help the lifting team to learn about the environment. The planning process is conducted by a team consisting of a lifting supervisor and one or more lifting managers. The investigation data and results are used to decide the best cranes, schedules, paths to conduct the heavy lifting. Details like equipment transportation and crane setup are also planned in this process. Finally, the execution team comprising drivers, signal men, riggers, and other workers execute the lifting job following the lifting plan decided in the planning phase.

The crane operations are highly dependent on knowledge and experiences. Among the lifting personnel, the signal men are responsible for giving operation signals such as swing, luffing, hoisting, and speed controls according to the plan. The signal men are also responsible for observing possible hazards during the lifting process such as collisions, overloading, and bad weather conditions. Crane operators have to manage the detailed operation of the cranes. Meanwhile, they are also required to learn about conventions of how to conduct safe and effective lifting jobs. These conventions are mostly based on the safety code as well as the experience of the crane operators. For example, the operators need to understand the load chart of the cranes and estimation of the pressures applied on the ground from the belts or outriggers in order to prevent the crane from crashing or tiling. On the other hand, cooperation within the lifting team involves a lot complex tasks such as plans reading, signals communication, decision-making, and problem troubleshooting.

Crane accidents may happen due to various reasons. Inadequate planning, lack of knowledge, bad communication, as well as machine errors and human errors may all lead to severe accidents causing unpredictable financial loss, equipment damage, time delay, and even death [9, 19]. Many accidents reported in OSAH are caused by misconduct due to lack of knowledge on safe codes, load charts, ground pressure calculations, or communication errors. Vocational training for heavy crane operators is of great importance to significantly reduce the safety risks and hence improve the productivity of heavy lifting.

1.2 Computer Simulation for Vocational Training

Crane lifting training usually has theory part and practice part. It is ideal to use actual cranes in practical training. However, most of the crane training centers often have only selected real cranes of representative models which are a small portion of a very large number of crane products in the market. Trainees typical will have several hands-on sessions once they pass their theory examinations. Due to the limited timeslots and also limited number of real cranes available, trainees will have to wait through a long queue before they can have their practical sessions. While training with actual cranes is necessary, just like in flight training, computer-simulated training can play a complimentary role in providing lifting training with several advantages. Virtual cranes of various models can be built in the simulation. Computer simulations also provide safe and cost-effective training for crane operations. With the virtual reality (VR) technologies, trainees are able to develop realistic experiences in computer-simulated training systems.

Computer simulation is widely used in vocational training. In the astronomy domain, NASA Ames Research Center launched the Vertical Motion Simulator [14] for astronaut training to reduce the risks in the space shuttle program. In transportation industry, Strayer and Drews [18] evaluated the benefits of simulator training on CDL truck drivers using the TranSimVS Truck driving simulator [10]. They designed a 2-hour training course for optimizing fuel efficiency through minimizing shifting and conducted on a group of drivers. The results which are measured in the following 6 months reported over 7 % improvements on unskilled drivers. In the domain of crane operations, a crane simulator using PC clusters was developed by Huang and Gau [7]. The simulator relies on local area network to communicate between independent modules. The system considers dynamics of the crane such as oscillations, collisions, and terrain following. Rouvinen et al. [17] presented a container gantry crane simulator for operator training based on the dynamic model, electric driving model, and collision detection. In 2014, Liebherr presented the LiSIM simulator for deep foundation machinery and crawler cranes. The LR 1300 crawler crane is modeled and used in the system to train the crawler crane operators in virtual environments [12].

Existing crane simulation systems are mostly designed for crane operational training of operators, signalmen, rigger, etc. Often, simplified plants or sites are pre-modeled for the training use. While general training is useful, functional training is getting more and more important especially when dealing with complex plants or sites, and complicated lift tasks. For this, traditional training methods are not sufficient when intelligence is highly desired to handle complex lift tasks. We are interested to design a simulation system for the training purpose combining general operation and complex operation. Evaluation will also be developed with the system. Hopefully, the system will offer an innovative solution for trainees to develop their vocational skills and build their confidence for crane operations.

2 Training System Architecture

To tackle the above-mentioned challenges in lifting work, we propose a vocational training system for heavy crane operations (Fig. 1). The system is built on digital crane and plant data with enabling technologies like VR, computer visualization, and artificial intelligence. On top of the data and technologies, the system also has customized design for vocational training of heavy crane operations. Three modules are designed for three different types of trainees (Fig. 2). The system accepts multiple types of digital plant data. They can be either in CAD or smart plant (e.g. PDMS) formats. For complex plants and those who have dynamic features, the system would rely on the point cloud data acquired from high definition laser scanning.

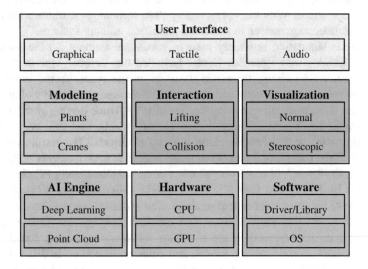

Fig. 1 Major components supporting the proposed vocational training system

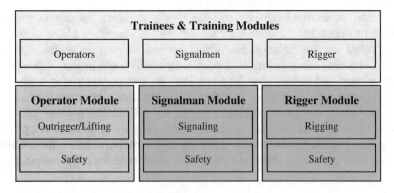

Fig. 2 Modules of the proposed crane lifting training system

Modeling technologies provide digital data of the cranes, plants, and crane accessories like hooks, rigs, and shackles. We have developed real-time rendering techniques to visualize the data on supporting various types of displays [1]. Realistic stereographics visualization is implemented with the training system. Users can interact with the system in several ways. Other than traditional keyboard–mouse inputs, trainees can learn lifting via touch screen, tactile devices, and voice navigation to interact with the training system. There are three types of interactions considered in the system: (1) the human–crane interaction for crane operating; (2) the crane–plant interaction mostly for the proximity and collision detection between the crane and plant, (3) the human–human interactions for communications, synchronizations, and cooperation between supervisors, drivers, operators, riggers, and signal men. Intelligence incorporated in the training system is enabled by the motion planning and VR technology [11]. Lifting path planning provides safe and optimized suggestions for the crane operators [2, 3]. Driving trajectory planning [20] suggests the crane driver with optimized turning trajectories. As the system is required to be real time, the Graphic Processing Units (GPUs) [16] are used to speedup the computation, especially for the collision detection and path planning modules.

3 Enabling Technologies for Simulation-Based Lifting Training

3.1 Modeling

The modeling process includes designing, reconstructing, or extracting data of the cranes, plants, and accessories. There are a large number of different designs of cranes existing. It would take enormous time to design the whole fleet of cranes. Our approach is to utilize the idea of product family modeling (Fig. 3). Cranes produced by companies like Liebherr and Terex have categorized into product families or series. Cranes in the same series share similar types of components. During the modeling process for cranes, standard components of the product family can be designed uniformly and adjusted in detail when assembled to a specific model.

Many modern plants have its digital version in PDMS format (Fig. 4a) which keeps all the parametric geometries of the plant structures. The training system is also able to accept Smart Plant format which is similar to PDMS. Beside, our system is able to accept CAD modeled plants.

To deal with old plants without any digital version, we propose to use the laser scanning technology (Fig. 4b) to capture the geometric objects in point cloud data. This is to create a discrete version of the plants using information including coordinate, normal vector, and color of each vertex captured. To capture the entire plant, multiple scans are often required. As the multiple scans may have their own coordinate systems, a stitching process is desired to fuse the scans forming a complete point cloud. The data after stitching can be very large leading subsequent

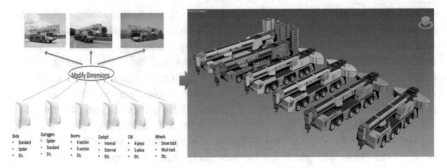

Fig. 3 The product family concept for modeling digital cranes

(a)   **(b)**

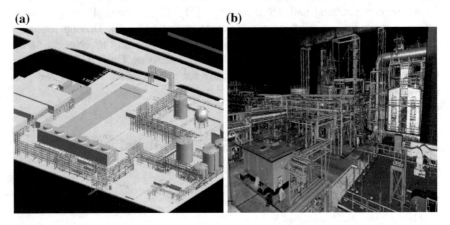

Fig. 4 Plant model inputs used in the proposed vocational training system. **a** PDMS plant. **b** Point cloud plant

jobs such as storing, retrieving, and processing to become difficult if not impossible. We have been developing several postprocessing techniques for data compression including downsampling [13] and resampling. Another promising technique is intelligent feature recognition [6] which can be used to reduce the size of scanned data by merging points in a same plane structure or a same pipe structure. This feature extraction process can be used to reconstruct 3D plant models. Particularly, we are interested to extract the road features from point cloud data for the purpose of mobile driving training. Figure. 5a shows scanned data and Fig. 5b is the extracted road feature from the point cloud.

3.2   Interaction

In this work, the proposed training system for heavy crane operators includes three types of interactions: human–crane, human–human, and crane–plant interactions. In the human–crane interaction mode, trainee operates on the virtual crane via a touch

(a)   **(b)**

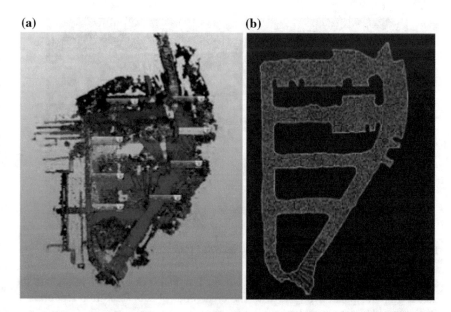

Fig. 5 **a** Multiple scans fused. **b** Road feature extracted

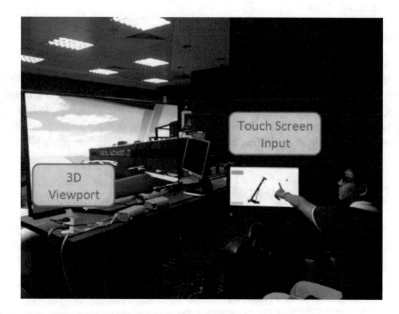

Fig. 6 The touch screen GUI for a trainee to operate a virtual crane

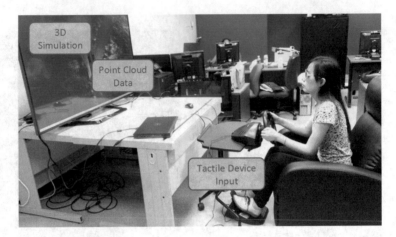

Fig. 7 Driving a virtual mobile crane in a point cloud plant using a tactile user interface

screen GUI (Fig. 6), or tactile steering wheel. The GUI is designed in relation to the actual graphics panel available with the crane control cab. Real-time rendering is implemented to give visual response for the trainee to obtain a realistic experience of crane operation. Figure 7 shows the tactile interface for a trainee to learn driving of a mobile crane using a steering wheel, brakes, and joysticks.

3.3 *Visualization*

Visualization plays an important role in simulated training. Trainees can opt for normal 3D or stereoscopic 3D during the training. To mimic the crane operation in the control room, the touch screen-based GUI is designed in normal 3D. However, trainees can choose to use the stereoscopic 3D visualization for the better depth perception which is important especially when dealing with heavy lifting in a highly complex plant.

4 Crane Operation Training

Our simulation system has several modules developed. In this section, we will discuss crane drive training, crane operator training, signalman training, and rigger training.

4.1 *Crane Driving Training*

Cranes as long vehicles are not easy to drive. Using the simulation system developed, trainees will be able to learn long vehicle driving in plants which can be modeled in PDMS or captured in point cloud form. Figure 7 shows that the trainee can use tactile devices such as steering wheel and joystick to mimic the driving in a 3D plant environment. In particular, our simulation system is able to detect collision which is important for safety driving [4].

4.2 Crane Operator Training

Figure 8 shows a crane control cab with a touch screen panel and front display. The front display produces synthesized view of the plant and the trainee can do the crane lifting control using the touch screen control panel. The lifting operation implemented includes boom extending, luffing, and rotating, as well as load hoisting. For safety training, trainee operators will receive proximity warnings when the crane and/or the lifting load are too close to the surrounding.

4.3 Crane Signalman and Rigger Training

The crane training system designed can be used for signalmen and riggers in addition to crane operators. Lifting is actually a collaborative job involving the whole lifting team. Lifting supervisors manage the lifting jobs. When mobile cranes are deployed in plants, crane operators will control the cranes from the control cab.

Fig. 8 Lifting training in a crane control cab with a touch screen panel and the trainee can view the plant from display panel

Fig. 9 Different views designed for different trainees

Riggers will attach the load with slings and hooks before the lifting, and detach the load when reaching the destination. Riggers will help to keep the load in a safe distance from the surrounding by applying suitable ropes connected to the load in operation. Signalmen, however, send information regarding the load such as possible operation (hoisting, luffing, swing, and load rotating) based on its current location and collision avoidance. Figure 9 shows the different views designed for signalmen, riggers, and operators. The signalmen will be alerted when the cranes and/or the lifting loads are within certain distance causing potential collision. Similarly, the riggers will be prompted to assist the moving of the loads at the beginning or at the end of the lifting task. These features are provided by the simulation system for safety training purpose.

4.4 Intelligent Crane Operation Training

Crane lifting especially in highly complex plants is a challenging task requiring a lot of trainings and retraining. The simulation system is able to provide optimal trajectory for crane driving (Fig. 10). This is extremely useful for large cranes to move in narrow roads.

The lifting path planning is a challenging problem. We have developed an innovative and real-time solution [2, 3] based on the Master–Slave Parallel Genetic Algorithms (MSPGAs) [5, 8, 15] (Fig. 11). The proposed system uses a fitness function to measure the energy cost, time cost, and safety cost of candidate paths initially generated by the initialization process. The algorithm relies on a fixed size population to evolve in the reproductive iterations which undergoes reproductive operators such as selection, crossover, and mutation during each round. Our algorithm MSPGAs use reproductive operators parallelized in GPU. GPUs have

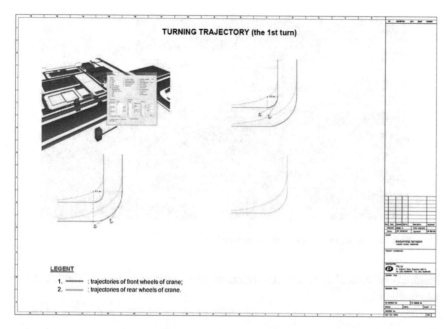

Fig. 10 Turning trajectory for intelligent driving training

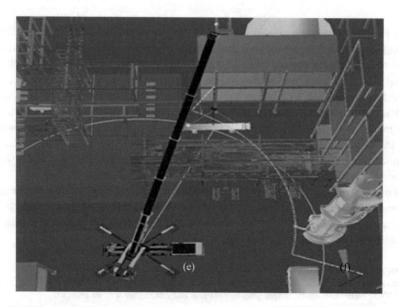

Fig. 11 Lifting path (*blue*—hoisting, *green*—luffing, *yellow*—swing, and *brown*—load rotating)

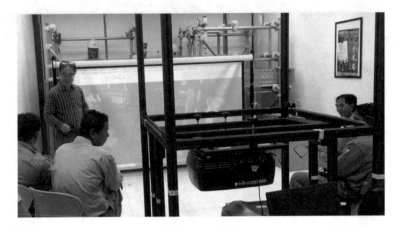

Fig. 12 A pilot training session with the simulation system

hundreds of streaming processors which allow a massive number of threads to run simultaneously.

5 Conclusion

The chapter presents a simulation system for vocational training of crane lifting team to improve their operational skills, safety awareness, and cooperation abilities. The system utilizes various types of digital models such as CAD, Smart Plant, and Point Cloud. Techniques supporting modeling, visualization, and interaction are described which can be used to help the lifting team improving their skills. Both general and intelligent operation training are available for the trainees taking the advantage of system for effective and efficient training. The system is user-friendly for training with touch screen, tactile devices, and voice navigations incorporated. A pilot training session was conducted in a lifting company on May 2015 which showed promising result (Fig. 12).

Future work includes a possible integration with existing training curriculum currently implemented in crane training centers or vocational schools. More safety codes on crane lifting will also be investigated for a possible integration with the simulation system for both training and assessment.

Acknowledgments PEC Ltd and Nanyang Technological University support this project. The authors would like to thank Mr. Huang Li Hui, Mr. Yang Bianyue, Dr Li Qing, Mr. Gong Yi, Mr. Lim Teng Sam, and many others who helped this project in one way or another.

References

1. Akenine-Möller, T., et al.: Real-Time Rendering. CRC Press (2008)
2. Cai, P., et al.: A GPU-enabled parallel genetic algorithm for path planning. In: 2013 Symposium on GPU Computing and Applications. Nanyang Technological University, Singapore (2013)
3. Cai, P., Cai, Y., Chandrasekaran, I., Zheng, J.M.: Parallel genetic algorithm based automatic path planning for crane lifting in complex environments. Autom. Constr. **62**(2), 133–147 (2016)
4. Chen, Y., et al.: Long vehicle turning. In: Simulations, Serious Games and Their Applications, pp. 85–103. Springer, Berlin (2014)
5. Fujimoto, N., Tsutsui, S.: Parallelizing a Genetic Operator for GPUs (2013)
6. Gumhold, S., et al.: Feature extraction from point clouds. In: Proceedings of 10th International Meshing Roundtable. Citeseer (2001)
7. Huang, J.-Y., Gau, C.-Y.: A PC cluster high-fidelity mobile crane simulator. Tamkang J. Sci. Eng. **5**(1), 7–20 (2002)
8. Ismail, M.A.: Parallel genetic algorithms (PGAs): master slave paradigm approach using MPI. E-Tech 2004, IEEE (2004)
9. King, R.A.: Analysis of crane and lifting accidents in North America from 2004 to 2010. Massachusetts Institute of Technology (2012)
10. L-3: TranSimVS Truck driving simulator (2015)
11. Latombe, J.-C.: Robot Motion Planning, Chapter (1996)
12. Liebherr: Liebherr presents its first Simulator for Operator Training on Deep Foundation Equipment at Conexpo (2014)
13. Moenning, C., Dodgson, N.A.: Intrinsic point cloud simplification. In: Proceedings of 14th GrahiCon, vol. 14, p. 23 (2004)
14. NASA: VERTICAL MOTION SIMULATOR. Retrieved 23 Mar 2015, from http://www.aviationsystems.arc.nasa.gov/facilities/vms/index.shtml#overview (2015)
15. Nowostawski, M., Poli, R.: Parallel genetic algorithm taxonomy. Knowledge-based intelligent information engineering systems, 1999. In: Third International Conference, IEEE (1999)
16. Nvidia, C.: Demos and Games. From http://www.nvidia.com/object/GTX_400_games_demos.html
17. Rouvinen, A., et al.: Container gantry crane simulator for operator training. Proc. Inst. Mech. Eng. Part K J. Multi Body Dyn. **219**(4), 325–336 (2005)
18. Strayer, D.L., Drews, F.A.: Simulator training improves driver efficiency: transfer from the simulator to the real world. In: Second International Driving Symposium on Human Factors in Driver Assessment, Training and Vehicle Design (2003)
19. Tam, V.W., Fung, I.W.: Tower crane safety in the construction industry: a Hong Kong study. Saf. Sci. **49**(2), 208–215 (2011)
20. Wang, D., Qi, F.: Trajectory planning for a four-wheel-steering vehicle. ICRA (2001)

Formative Evaluation of a Mathematics Game for Out-of-School Children in Sudan

Hester Stubbé, Aiman Badri, Rebecca Telford, Stefan Oosterbeek and Anja van der Hulst

Abstract Education for children in the developing world is in crisis, and children growing up in war are already at the sharp end of global development challenges: of the 57 million primary-age children who are out of school, almost half live in conflict zones (http://www.unicef.org/media/media_75652.html). This research is grounded in the issues of educating children living in these zones, with a particular focus on innovative approaches to access via online and distance learning using ICT. Within this approach, a mathematics game was developed in close cooperation with the Ministry of Education in Sudan, the Ahfad University for Women in Khartoum and children from target communities. The intention was to develop a game that enables children from those communities to autonomously learn mathematics covering a curriculum of 3 years. The objective of this research is to describe the design of this game in relation to its requirements—educational, contextual, cultural, and from a game design perspective. An additional goal is to find out if children can play the game, want to play it, and will play it again. For pragmatic reasons this evaluation has been carried out in the Netherlands, with a total of 15 Arabic-speaking children. Their backgrounds were Egyptian, Moroccan, and Sudanese. The ages varied between 5 and 9, and the group was made up of seven girls and eight boys. The game was played on three consecutive Saturdays;

H. Stubbé (✉) · A. van der Hulst
TNO, Soesterberg, The Netherlands
e-mail: hester.stubbe@tno.nl

A. van der Hulst
e-mail: anja.vanderhulst@tno.nl

A. Badri
Ahfad University for Women and UNESCO, Khartoum, Sudan
e-mail: uscasd@hotmail.com

R. Telford
War Child Holland, Amsterdam, The Netherlands
e-mail: rebecca.telford@warchild.nl

S. Oosterbeek
University of Amsterdam, Amsterdam, The Netherlands
e-mail: stefan.oosterbeek@gmail.com

© Springer Science+Business Media Singapore 2017
Y. Cai et al. (eds.), *Simulation and Serious Games for Education*,
Gaming Media and Social Effects, DOI 10.1007/978-981-10-0861-0_5

61

most children (8) played the game three times, some children played the game once (2) or twice (5). Using observation forms, semi-structured interviews—with the children and with their parents—and questionnaires, data about motivation, enjoyment, perceived difficulty and cooperation between children were collected. The game met most of the requirements. Educationally, the game taught the mathematical concepts in a way that was understood by the children. This means it fits the context as well: children can learn autonomously. Culturally, we received anecdotal evidence that the Sudanese context was portrayed in a correct way. From a game design perspective, there should have been more learner control, though. The results of the evaluation show that average enjoyment increased slightly over the 3 weeks: from 3.5 on a 5-point Likert scale in week 1 to 3.9 in week 3. The average perceived difficulty was low: starting at 2.6 in week 1 and decreasing in week 2 and 3 to 1.7. Average motivation was high: 4.5 on a 5-point Likert scale, and stayed almost the same during the evaluation. Average cooperation was high: around 4.5 on a 5-point Likert scale. The setting of this evaluation did not completely reflect the situation in Sudan; the children lived in the Netherlands and knew their mathematics. Still, it has provided worthwhile feedback to improve the game further.

Keywords Game-based learning · Mathematics · Primary education · Developing countries · Out-of-school children · Game design

1 Introduction

Education for children in the developing world is in crisis. Children growing up in complex emergencies are at the sharp end of global development challenges: of the 58 million out-of-school children globally, 36 % live in countries scarred by war and violence. More than one-third of refugee children globally are missing out on primary education [34] and the safety and education of girls are disproportionately affected [17]. Reaching those without access to school is a pressing issue, particularly girls and children in the rural areas in Africa and South Asia [4, 9, 19]. Complex and mutually reinforcing patterns of disadvantage—poverty, gender inequity, disability, conflict, and displacement—raise barriers to schooling and erode educational opportunities for children.

In recent years, global policy frameworks on education reflect the "*shift in the global conversation on education from a focus on access to access plus learning*" [32]. The proposed Sustainable Development Goal (SDG) on education along with the Incheon Declaration adopted at the World Education Forum in 2015, outline a vision for sustainable and equitable education spanning preprimary education through to at least lower secondary education. There is also recognition of the need to focus on quality as well as on access, and to support the most disadvantaged children, not least those living in emergencies. There are not enough qualified teachers, or support for teacher professional development; class sizes are large and underresourced; and classroom methods are teacher-centered. This leads to high dropout levels, up to 50 % [29].

As other sub-Saharan countries, Sudan struggles with the same issues [30]. Enrollment in basic education has increased to 80 %, especially since 2005, as a result of peace. This still means that approximately 1 million children have never been to school. Dropout rates are also relatively high: whilst there is a fairly high intake of 80 % into grade 1, completion rate is only 54 %. Suggested reasons for this are the costs of schooling in relation to perceived benefits: '...the quality of schooling is too low to justify the student's time and the direct costs in terms of parental financial contributions or the opportunities of a child who could otherwise be contributing to the household income or helping with chores' [30, p. 4]. Other factors involve the students' travel time to school; when schools do not offer all grades of basic education—as is the case for village schools and for children from pastoralist communities—children need to change schools to continue their education which may lead to their dropping out. There are significant disparities with respect to region, rural–urban location, gender, and income. "Children in rural areas, those from poorer households and girls are at a disadvantage. The strongest predictor of access to schooling is whether a child lives in an urban or rural area, with urban children being 17 % more likely to have access to education" [30, p. 8]. In addition to this, vulnerable populations like nomads and internally displaced persons (IDP) have little access to education. Among these vulnerable groups, the share of girls in basic education is even smaller than that in regular schools (38 % girls in nomadic schools compared to 44 % in IDP schools and 47 % in regular government schools). Lack of safety on the road to school and in the school itself is mentioned as one of the reasons [30, p. 9].

At present, it is not realistic to believe that issues around access to quality education for all will be solved through traditional means. Due to the cultural, geographic, and socioeconomic background of children, more flexible, empowering, and affordable approaches, outside formal schools are required. Online and distance learning with ICT are seen as possible solutions. The focus of this approach should be on rural areas, communities affected by conflict—including reaching IDPs—and specifically include girls and minority groups. Consequently, the overall question that needs to be answered is: how can we support children to learn autonomously for a longer period of time in the rural, sub-Saharan post-conflict context? Because children will need to learn without a teacher, for a longer period of time, we chose to develop an educational game in order to teach and motivate children to learn at the same time. This chapter will focus on the requirements for successful learning for out-of-school children in Sudan and how these requirements can be translated into a working game design. Although the children will need to learn how to read and write as well, for pragmatic reasons we chose to start with mathematics.

The first part of this chapter describes the requirements for successful learning for out-of-school children in the rural areas in Sudan. The second part is a description of the design of the game. The last part of this chapter is a formative evaluation of the game with 15 Arabic-speaking children in the Netherlands, in which we investigated whether children can play the game, like to play it, and want to play it again. Results from this formative evaluation will be used to improve the game further.

2 e-Learning in Developing Countries

To allow children to learn autonomously, the use of e-Learning was perceived to be the most likely option to enhance access to education. Although e-Learning in developing countries is a field of interest and scrutiny both by researchers and educators, evaluation has focused on provision of hardware (e.g., number of laptops distributed) and anecdotal evidence of specific 'successes'—often self-reported [14, 15, 39]. It is generally accepted that educational technology in itself cannot enhance learning in developing countries [24]. To date there is some evidence from developing countries on how educational technology can support learning. However, there are indications that findings in developed countries may not be applied to developing countries [35].

There are many well-known examples of unsuccessful e-Learning schemes in Africa [2]. Unwin [36] outlines significant acknowledged challenges of successfully implementing an e-Learning program, or any technology for development program, within a developing country context. Nonetheless, there are a few notable examples of open access and distance learning for early literacy, basic and primary education as useful tools in bridging the gaps of access due to conflict, gender, and geography [33, 38]. The UNICEF report, looking specifically into the effectiveness of the learning materials, and less into the cost-effectiveness and scalability, cited three key factors to reaching children who were currently excluded from formal education: (1) location, (2) flexibility, and (3) continuity. Specific needs identified were (1) an appropriate location for accessing learning materials and supplementary face-to-face contact, (2) flexibility in learning alongside other demands of the family, which might interrupt a traditional school schedule, and (3) the opportunity for progression into the mainstream educational system if desired by the learner and their family. This is in line with a more general conclusion of Clarke [6, p. 1] 'in order for technology to improve learning, it must "fit" into students' lives… not the other way around'.

3 e-Learning and Games

Although the effectiveness of e-Learning and games in developed countries may differ from the effectiveness in developing countries, the existing evidence on the effectiveness of e-Learning and games within the context of developed countries formed the basis of this research. In their meta-review on the effect of e-Learning and educational games, Vogel et al. [37] found an overall positive effect: significant higher cognitive gains were observed in subjects utilizing interactive simulations or games versus traditional teaching methods. Subjects using educational technology also had a more positive attitude toward learning than those in the more traditional educational settings. This seems to be the case for boys as well as girls, although the low number of studies gives reason to consider these results with caution. All age groups showed significant positive results for the use of educational technology. The type of activity did not appear to be influential; neither did realism of the pictures in the game. Both individual users as well as groups showed significant positive results. Programs that were designed to automatically navigate students through the system seemed to be less effective than traditional classroom education, but the sample size made it impossible to draw this conclusion with confidence.

Effective use of educational technology is characterized by a clear focus on curriculum, teacher development, and evaluation [24]. As children in remote areas will have to learn without a teacher, this automatically places the curriculum and the learners at the center of attention, with technology as an enabler to achieve learning [26]. The pedagogic framework around this is based on active learners, and stresses the activities and interaction of learners, instead of content in the sense of preprepared learning materials [8, 16]. The activities themselves are central in this pedagogical vision.

4 Requirements

It is the intention to develop an educational game that supports children to master the first 3 years of the Sudanese mathematics curriculum for out-of-school children, autonomously. After 3 years of primary education, there is an official mid-basic education exam in Sudan. The educational game will prepare children to take this exam.

Identifying the requirements for an educational game in a developing country is complex: not only does it require the input from different domains of expertise—development context, education, learning with ICTs and the subject domain—it also needs an integration of all these different types of input. In this paragraph, these domains of expertise will be related to the context in Sudan. This leads to a set of requirements for the game per domain of expertise. The final set of requirements may not be the best for each of these domains separately, but will be the optimal set in this specific context.

4.1 Development Context

Context:

Research [2, 33, 36] shows that educational programs with or without the use of ICT have not always been successful in developing countries. The three most important factors for an educational program to be successful in developing countries are: (1) location, (2) flexibility, and (3) continuity. In addition to this, the target population of game is rather specific: children between the ages of 6–11, boys, and girls. These children have never been to school and can be assumed to be illiterate. Their parents are probably functionally illiterate as well. Finally, Sudan is an Islamic country. The narrative, graphics, and examples used should not be offensive in this respect (e.g., showing the soles of shoes).

Requirements:

- Location: The children will learn in their own communities. As a result they will have to learn without a teacher, because teachers are not available in the villages. This means instruction will have to be embedded in the game. Facilitators need to be present to motivate learning and help with technical problems.
- Flexibility: Learning will take 2 hours a day at the most. The official curriculum for out-of-school children in Sudan comprises three subjects: mathematics, reading and writing and religious studies, a combination of history, biology, and religion. Forty-five minutes will be spent on mathematics, 45 min on reading and writing, and 30 min spent on religious studies. The most important rationale behind this is that these hours can be fitted around additional responsibilities such as caring for family members or doing chores. In addition, hospital education is based on similar amounts of time spent on basic learning (mathematics, reading, and writing).
- Continuity: The official Sudanese curriculum should be used. Children should be able to pass the official exam taken after 3 years of primary education. In the meantime, progress of individual children must be logged and stored in a management system. In addition to that, the Ministry of Education in Sudan and the National Councils for Adult Education and Illiteracy and for Mathematics should be involved in the development of the game.
- Age: The game must be appropriate, appealing, and motivating for children between the ages of 6 and 11.
- Gender: The game must be appropriate, appealing, and motivating for both boys and girls.
- Illiteracy: The game should not require the children to be able to read to understand instruction.
- Culturally appropriate: The game concepts, graphics, and examples of mathematical problems must be culturally appropriate.

4.2   Education

Context:

As the children are illiterate and will have little learning support from parents and teachers, we have to assume they have little to no informal mathematical knowledge (e.g., numbers or quantities). In addition, we will make use of known effective methods of teaching.

Requirements:

Conditional knowledge: Because the children will probably lack informal mathematical knowledge, it is best to use the approach for 'struggling learners' [13]. The level of informal knowledge influences the capacity to understand instruction, which means that instruction must be very clear for children to be able to understand. This means that the game offers direct instruction [23, 31]: new concepts are introduced using short and clear explanations. This should be done in the mother tongue, to make sure that children do understand the instruction. Children should be allowed to repeat instruction as many times as wanted or necessary.

Prior knowledge: The curriculum of the game should start from the very beginning, with numeracy and oral counting. Some children will be able to progress faster through the initial stages of the game, but no child will be frustrated because it does not understand what to do.

Mastery learning: All children can learn to do mathematics, but some need more time than others. The game should only allow children to continue with new concepts of mathematics once they have mastered the previous one. This can be measured on the basis of the number of mistakes within a certain time frame. Children that master a concept should be able to progress more quickly.

Time on task: Time on task is a strong predictor of learning results [20]. Children should be actively involved in many different minigames. This increases the number of exercises they do, and contributes to the time spent on learning. In this way, children can repeat and practice new concepts many times, without having the feeling that it is repetitive [18].

Reproduction and production: For children to master a new concept, they will have to be able to do reproductive and productive exercises. The minigames should provide several types of activities ranging from a 'multiple choice' activity where children choose the right answer, to 'matching' and 'arranging' numbers and amount and finally 'writing' the correct answers to problems on the tablet.

Acquisition of concepts and skills: the minigames should not only provide opportunity to improve skills (e.g., proficiency in calculations), but also focus on the acquisition of mathematical concepts, which is based on problem solving [10].

4.3 Learning with ICTs

Context:

Research has shown that ICT has a positive effect on learning mathematics, especially for younger children [7, 28]. Three important characteristics of learning with technology for young children are (1) interactivity; the active participating of the child, and freedom to influence the order and pace. (2) Functionality; the right balance between content and presentation (neither too dull because of the abundance of content, nor too playful or with an overly 'flashy' presentation). (3) adaptive feedback; feedback on results as well as on process [21, 27]. In addition, the use of technology provides the opportunity to motivate children to learn for a longer period of time.

Requirements:

- Interactivity: The game should invite children to actively engage with the content. They should be allowed to do many different minigames, and choose— to a certain degree—what they do first.
- Functionality: The game should be rather simple and intuitive. The graphics should appeal, but not attract too much attention. There is no need for extra sound, apart from instruction and feedback.
- Adaptive feedback: There should be feedback on right and wrong answers and on progress.
- Support: If a child does not interact with the game for 1 min, there should be an audio fragment asking if the child needs any help. At all times, children should be allowed to ask for help on where to go next, or to repeat the instruction of the exercise.
- Motivating elements: The game should have motivational elements that engage children and create flow (children want to finish the next level). It should use several types of motivational aspects, like competition, competence, and cooperation.

4.4 The Subject Domain

Context:

As mentioned before, the curriculum used in the game is to be the official Sudanese curriculum for out-of-school children. This curriculum is very similar to other curricula for the first years of mathematics all over the world [3]. Still, it is too often assumed that mathematics is a universal language and that mathematics curricula are easily adapted [12]. Akkari [1] argues that the differences in the Arabic mathematics education are significant enough to warrant special caution, because like language it may contain sensitive issues imbued with symbolic political and cultural value. To avoid a psychological blockade, examples should be based on everyday life. Moreover, mathematics should not be used to implicitly teach political values, by denying the national character or imposing, e.g., colonial values. This would reinforce patterns of disadvantage and inequality. In addition, we will make use of known effective methods of teaching mathematics.

Requirements:

- Topics: The topics to be covered in the first 3 years math curriculum are: oral counting fluency, one-to-one correspondence (counting objects while pointing at each object that is counted), number identification, quantity discrimination, missing number, addition word problems, addition problems, subtraction word problems, subtraction problems, multiplication problems, division word problems, division problems, time, weight, length, and shape recognition [25].
- Cultural aspects: In the instruction of new concepts, the examples used and the exercises should reflect the cultural and geographical background of Sudanese children. Furthermore, specific care should be given to neutral examples: the example should not contain implicit discrimination or references to conflict. It is best to use everyday examples that children will recognize, like fruit and vegetables.
- Teaching mathematics: Effective teaching of mathematics should involve instruction on concrete examples, instruction on a model of reality and on abstract problems. Preferably, the concrete examples and the abstract problems are introduced and used at more or less the same time, in both instruction and practice. Therefore, all instructions should show concrete examples, the associated model, and the abstract problem, and the various exercises should also provide these three different types of practice.

Design of the Game

The three sets of requirements listed above, were integrated into a final set of requirements. Mostly, the requirements dealt with separate aspects of the game, which meant they could all be included. There was a conflict between the requirements 'interactivity' and 'mastery learning'. Where interactivity implies children have a certain level of control over order and pace, mastery learning states that new subjects should only be introduced when the previous one is mastered. This leaves the children with less control over their activities. In this first version of the game we have chosen to guide learning according to mastery learning principles, and allow children only limited control over their actions. This choice was supported by the fact that the Sudanese Ministry of Education preferred the curriculum to have a specific order.

In March 2012, during a week of meetings with the Ministry of Education in Sudan and the relevant National Councils, the curriculum for the game was agreed upon. During the development of the game, they approved and endorsed the versions of the game.

The educational game for mathematics was developed in three phases. The first version allowed for 6 weeks of learning and was tested in a small pilot in the period December 2012–February 2013. Based on the findings in this pilot, the second version was developed that allowed for 6 months of learning. This version is being tested in a large pilot in the period October 2014–March 2015. Based on the results

Fig. 1 Screenshot
gameworld 1

Fig. 2 Screenshot
gameworld 2

of this pilot, the final version of the game was delivered in June 2016, covering 3 years of learning. Following the 'fastest route to failure', we first tried out if children could learn from an educational game, without teachers (proof of concept). Only then did we develop the 3-year curriculum. The resulting game incorporates two distinct levels, each with a different pedagogy. The first level is that of Game Worlds which provide the connecting narratives for the second level, that of separate minigames (44 different minigames, 160 variations of minigames). Game World 1 (see Fig. 1) is about helping other children to achieve goals in their lives; by doing minigames, children help other children to become, e.g., a goat herder or doctor. Half of the jobs are familiar roles within the target communities, such as a cook, tractor owner, or brick maker. The other half are known to the children, but belong to the outside world, like a teacher, nurse, doctor, and engineer. In a playful way this helps the children to broaden their future perspective. Game World 2 (see Fig. 2) is a shop where children can buy and sell products. By playing the minigames, children can increase the number of products they can sell and enhance their shop.

In the second level, the minigames address specific mathematic concepts. Some minigames have variations that can be used for a number of mathematic concepts. All mathematic concepts can be practiced by a number of different minigames. This is supposed to help the children to understand the concept and stay motivated.

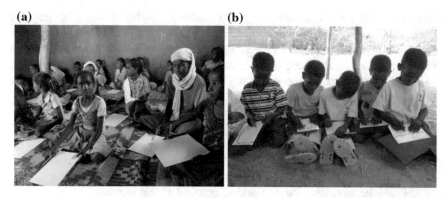

Fig. 3 **a** and **b** Children drawing their environment

Navigating the interface is easy and intuitive. Logging-in can be done by tapping your own photograph, and pictograms are used to navigate through the game. The game worlds are attractive and the minigames diverse and based on repeating the concepts without being repetitive. The graphics used are simple, but not childish. The colors are bright, but also realistic; the game worlds reflect the real world. Both boys and girls in remote villages in Sudan have provided input for the graphics, colors, and game worlds of the game. The graphics are based on drawings the children made themselves (see Fig. 3a, b), to make sure they would recognize them. These graphics were checked in Sudan before they were used in the game, by children and adults.

All instruction is in audio and video; the language used is formal, but easy Arabic. Instruction on how to use the game is given by a woman's voice. The instruction videos about the mathematic concepts were made with children, boys as well as girls. In this way we did not only provide simple instruction on mathematic concepts, but also introduced role models.

Several types of motivational aspects were included: (1) competence: children can see their progress and what they have learned already, (2) competition: there are many levels, and children can compare their level to that of others, and (3) cooperation: in the setting in which the game is used, children can help each other. Thus, we can motivate more children. Finally, there are some fun elements, like a football that can be kicked around, but the game is not too playful.

5 Formative Evaluation in Experimental Setting

An evaluation of the effectiveness of the game can only be done in Sudan, with out-of-school children. Still, we know that enjoyment, motivation, and perceived difficulty are important factors that influence learning outcomes: motivated children, who like to play the game and do not find it too difficult, will learn more than children who do not like the game or think it is very difficult (Flow theory by

Csikszentmihalyi and Hermanson [5]). To establish whether children can play the game, like to play it, and would like to play again, we carried out a formative evaluation with children enrolled in the Weekend School in The Hague (the Netherlands). Although these children are very different from the target group in Sudan, it allowed us to supervise playtesting and test general motivation with respect to playing the game.

5.1 Method

All children between the age of 5 and 9, enrolled in the Weekend School in The Hague to learn Arabic, played the game for three consecutive Saturdays, for one hour per day. They played in pairs, mostly for the pragmiatic reason that there were not enough tablets, but also to stimulate cooperation. During and after each playing session, data were collected on Enjoyment, Perceived difficulty, Motivation, and Cooperation. Data were collected with four different instruments: an observation form, a short questionnaire, a semi-structured interview for children, and a semi-structured interview for parents, specifically designed for this evaluation. During their play (see Fig. 4a, b), the children were observed according to the observation form. At the end of each playing session, each child was asked a few questions (semi-structured interview) and was asked to answer a short questionnaire. The answering categories are a 5-point scale, portrayed in five cups, ranging from completely empty to completely filled, to facilitate answering by young children. At the end of the third Saturday, parents were involved in a semi-structured interview about their children's experience and perceptions. Before the evaluation started, parents gave oral consent for their children to participate.

To ensure that the children would be able to understand the questionnaire, it was pretested with two children within the right age range. The children understood the questions well, and the pretest did not provide any input for further adjustments.

(a)   **(b)**

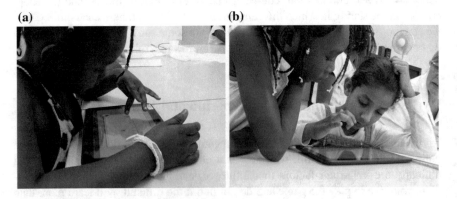

Fig. 4 a and **b** Children in The Hague playing the game

Fig. 5 Number of boys and girls per week

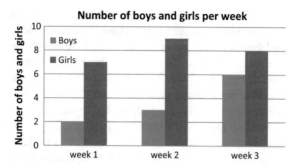

5.2 *Population*

The Weekend school started around ten years ago, and convenes every Saturday. Children come together to learn Arabic; parents meet as well. A total of 15 children participated in the evaluation, seven girls and eight boys, aged between 5 and 9. For an overview of the number of boys and girls per week, see Fig. 5). Originally, they are from different Arabic-speaking countries (Sudan, Egypt, and Morocco), but they have all been in the Netherlands for all or most of their lives. Obviously, they are all enrolled in formal schools in the Netherlands and know their mathematics. As learning results were not evaluated, children were not tested for their level of mathematical knowledge.

Not all children were present every Saturday (see Fig. 5). Still we decided to include all data, because it provides a valuable insight in how children play the game. We only excluded the data of a 5-year-old girl because her answers on the questionnaire were found to be unreliable: all answers were in one category (the highest category).

5.3 *Results*

The observations showed that children were motivated to play the game. They enjoyed the minigames and experienced competition between the pairs ('How much have you completed? We have two stars already!'). At the same time the children were sometimes frustrated by minigames they did not understand immediately. One minigame was not working properly, and all children were frustrated about that. We also observed that children passed the tablet to other children once there was a tricky minigame. One girl quickly found out that she was allowed one mistake per minigame. Without trying to give a correct answer, she just made the mistake, 'so they could be very fast in finishing the minigame'. One child asked for silence. The children were sitting rather close to each other, and there was quite a bit of noise. This child preferred to work in silence, and alone.

The test version of the game still had English subtitles, to facilitate demo sessions and improvement of the game; the subtitles allow non-Arabic speakers to play and give feedback on the game as well. Although the children were advised not to use these, of course they did read them, and sometimes found the right answers to exercises. One of the fun features of the game was that children could kick a football. None of the children found this feature by themselves. When they were made aware of this, they chose not to use it: they wanted to finish their level! Overall progress was shown in a number of stars and in reaching the next level of the game. Boys tended to have more interest in the number of stars they had, girls were more interested in the beauty of their hut (flowers) or an extra little baby goat for the goat herder.

The semi-structured interviews with the children did not really provide any extra information. Children expressed that they liked the game, wanted to play on, and would like to play again the next week.

The results of the questionnaire were analyzed according to the constructs— Enjoyment, Difficulty, Motivation, and Cooperation. To do this, some of the scores on Difficulty had to be recoded because they were phrased in a positive way. Then averages of the constructs were calculated (see Fig. 6).

Because of the small number of participants, no extra analyses were performed for age and gender, and no statistical analyses were performed.

Enjoyment was 3.6 in week 1 and increased to 3.8 in week 2 and 3.9 in week 3. Difficulty was 2.5 in the first week and decreased in week 2 (1.7) and slightly increased again in week 3 (1.8). Motivation was high (4.7) in the first week and stayed high over the weeks (4.75 in week 2 and 4.7 in week 3). Cooperation was high in week 1 (4.8) and slightly decreased over the weeks (4.7 in week 2 and 4.5 in week 3).

The semi-structured interviews with the parents confirmed the results of the children's semi-structured interviews: the children enjoyed playing the game, were motivated to do well, and were looking forward to playing again. One of the parents

Fig. 6 Average score per construct per week

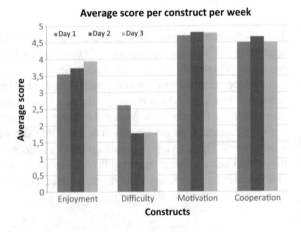

said that the children had learned in three different ways: they had practiced their Arabic, their mathematics, and at the same time learned about their parents' home country because of the graphics in the game.

6 Discussion and Conclusion

In the study reported in this chapter, we aimed to develop an educational game that allows out-of-school children in Sudan to learn the first 3 years of mathematics of the Sudanese curriculum, autonomously, and evaluate if children can play the game, like to play it, and want to play it again. Although there are many e-Learning programs that teach mathematics, the introduction of them in developing countries has failed for various reasons [2]. The three most important conditions for successful education in developing countries are locality, flexibility, and continuity. This means that learning should take place in the communities where the children live, take a maximum of 2 hours a day, and lead to an official certificate. As a result, the children will have to learn without a teacher because there are no schools in the communities. This leads to very specific requirements for the game regarding instruction, feedback, and motivational elements. Most existing mathematic applications or e-Learning programs only provide the opportunity to improve skills and do not focus on instruction (Rekentuin; [22]). This makes them unsuitable for learning without a teacher. Another issue with the existing materials is that the language and examples used do not match the cultural context. Finally, motivational elements do not always support learning autonomously for a longer period of time.

In cocreation with all project partners, children as well, we developed the game in three phases. First proving this concept can work, then trying out with a larger group of children, over a longer period of time, and only then developing the full 3 years of curriculum. This approach should contribute to an optimal game design, including all important aspects from various perspectives. The Ministry of Education and National Councils in Sudan have approved and endorsed the game concept, the children recognize their own surroundings and from the perspective of education the game supports learning mathematics for this target population.

The game met most of the requirements. Educationally, the game taught the mathematical concepts in a way that was understood by the children, without extra instruction. This means it fits the context as well: children can learn autonomously. Culturally, we received anecdotal evidence that the Sudanese context was portrayed in a correct way. From a game design perspective, there should have been more learner control, though.

In general, the various instruments yielded similar results; the data of the different instruments support each other. This makes the results more reliable. Overall, children liked playing the game, knew what they had to do, enjoyed playing together and were motivated to play again. The fact that they knew what to do and found the game not too difficult is no guarantee that the instruction is perfect. These children go to school every day, and know their mathematics. On the other hand, they sometimes struggled with the Arabic, which made the game a little bit more difficult for them. The fact that they enjoyed the game, which motivated them, is a very positive finding. These children have access to television and entertainment gaming. They probably use some educational technology at school as well. Although this is a rather simple game, it does create flow and engages children, even those that have access to much more exciting entertainment games.

Still, there are some interesting discrepancies as well: children mentioned that they thought the game was not too difficult. The observations, however, showed that children did get frustrated and sometimes even passed the tablet on to others because they could not answer the exercise. This can be due to children giving socially desirable answers. Another reason might be that the overall impression of the game is that it is not too difficult, whereas some of the minigames may be difficult. The other discrepancy has to do with cooperation. Children mentioned that they enjoyed working together, but observations showed that they were more taking turns than working together. Answers were, in the end, given by the child whose turn it was. They did not discuss answers or strategies. Some children gave up and passed the tablet on and some children just gave the answer, even when it was not their turn. That created quite a bit of frustration with the child who needed a little more time to find the right answer, but was very motivated to do well. Also, one of the older boys asked if he could work by himself on the third Saturday; he did not want to cooperate with anybody else. It is possible that the feeling of competition between the pairs has had a negative effect on the element of cooperation. Children preferred to choose for the fastest way to success, i.e., next level. This could be by working alone, having somebody else answer, or just by shouting the right answer.

The evaluation did reveal a trade-off between direct instruction and mastery learning on one hand and the possibility for children to influence the order and pace of the game on the other hand. All children mentioned that they had no influence over the order of the minigames in the game. The current version of the game provides optimal instruction and guided learning paths, but at the same time takes away most learner control. When children tap the screen to continue, one minigame appears. They cannot choose. Although the children all felt the same, their motivation and enjoyment did not decrease during this short experiment.

The formative evaluation as carried out in the Weekend School in the Netherlands shows that even Arabic-speaking Dutch children enrolled in the Weekend School in The Hague to improve their Arabic, who have access to television and entertainment games, enjoyed the game and wanted to play again. The evaluation lasted for 3 weeks, and in this time Enjoyment, which was high from the start, increased slightly, Motivation stayed high, and Perceived difficulty was not high to start with and decreased after week 1. Cooperation decreased slightly. Because of the small number of participants a statistical analysis was not possible, these differences may be due to chance. Observations showed that children were frustrated from time to time, by minigames they did not understand quickly, or by a minigame that did not function properly. This frustration did not show in their scores on Motivation, Enjoyment, and Perceived difficulty.

The most important findings were that children were not really cooperating when playing in pairs, and that the game prescribes what children can do: they have no control over the order in which they do the minigames.

During the formative evaluation, children worked in pairs. The most important reason behind this was the number of tablets available for this evaluation: there were not enough tablets to play individually. Cooperation, which is described as a motivating factor in game-based learning, took the form of taking turns during the evaluation, passing the tablet on to others and have them answer the problem, or just shouting the right answer. The reason behind this might be the competition the children felt between the pairs. They wanted to be better than the others, and thus wanted to increase their speed. One boy even asked to play alone on the last Saturday; he felt he could progress faster when playing alone. Competition stood in the way of cooperation, and perhaps even in the way of learning for some of them. For now, in Sudan, the situation will be different: children will play the game individually. To assess their individual progress, they will need to do the minigames themselves, using their unique account. They can help each other or ask for help, because they will be playing the game in the same hut, at the same time. We expect some competition between the children, but this cannot have the same effect as in the evaluation. Children play individually, they cannot take turns, or pass difficult problems on to another child. In the end, they have to answer all problems themselves.

In the questionnaire, all children answered that they had no control over the order in which they played the minigames. Indeed, the version of the game they played did not give them any control over this. They were in control over the pace in which they played the minigames, and they could watch the instruction videos as many times as they liked. Meta-analyses on the effectiveness of gaming show that the positive effect of educational technology is reduced by less learner control [11]. Since this game is meant to support autonomous learning for a longer period of time (3 years), this may become an issue. Also, motivation may decreases because of this. On the other hand, research on learner control is based on developed countries. Because of cultural differences, in developing countries, and in Sudan specifically, children are not used to having control. Others decide for them, and they follow. From this perspective, the lack of learner control may have less impact in sub-Saharan Africa, than it would in the Netherlands. Still, we will need to explore how we can add the element of learner control, without losing the mastery learning approach and the support that struggling learners need. This will help to improve the game.

Overall, the research-based design and cocreation has led to the development of a mathematics game that seems to meet most requirements and is acceptable and accepted by all stakeholders, including children. We will include more (illusion of) learner control, to make sure that the game stays motivating over a longer period of time. The results of the pilots in Sudan will show how effective the game really is.

Acknowledgments e-Learning Sudan is conceived through a collaboration between the Ahfad University for Women and War Child Holland. It is funded through the Conn@ct.Now program with additional resources from UNICEF Sudan. Curriculum, game development, and technical support was provided by TNO. The game was produced by Flavour, with support from creative partners in Sudan. e-Learning Sudan is presently hosted by the National Council for Literacy and Adult Education (NCLAE).

References

1. Akkari, A.: Education in the Middle East and North Africa: the current situation and future challenges. Int. Educ. J. **5**(2), 144–153 (2004)
2. Bitew, G.: Using plasma TV broadcasts in Ethiopian secondary schools: a brief survey. Aust. J. Educ. Technol. **24**(2), 150–167 (2008)
3. Bodovski, K., Farkas, G.: Mathematics growth in early elementary school: the roles of beginning knowledge, student engagement and instruction. Elementary Sch. J. **108**(2), 115–130 (2007)
4. Burnett, N.: Priorities and strategies for education—a world bank review: the process and the key messages. Int. J. Educ. Dev. **16**(3), 215–220 (1996)
5. Csikszentmihalyi, M., Hermanson, K.: Intrinsic motivation in museums: why does one want to learn? In: Hooper-Greenhill, E. (eds.) The Educational Role of the Museum, pp. 146–160 (1999)
6. Clarke, D.J.: e-Learning: big bang or steady evolution? Learning Technologies (2002)
7. Clements, D.H., Sarama, J.: Effects of a Preschool Mathematics Curriculum: Summative Research on the Building Blocks Project. The National Council of Teachers of Mathematics, Inc. www.nctm.org (2007)
8. Collis, B., & Moonen, J.: Pedagogy: making the U-turn. In: Flexible Learning in a Digital World. Open University, Netherlands (2001)
9. Cremin, P., Nakabugo, M.G.: Education, development and poverty reduction: a literature critique. Int. J. Educ. Dev. **32**(2012), 499–506 (2012)
10. Galen, van, F., Jonker, V., Wijers, M.: Designing Educational Mini-Games. ISDDE 2009 Cairns (2009)
11. Garris, R., Ahlers, R., Driskell, J.E.: Games, motivation, and learning: a research and practice model. Simul. Gaming **33**(4), 441–467 (2002)
12. Gerdes, P.: On culture, geometrical thinking And mathematics education. Educ. stud. Math. **19** 137–162 (1988)
13. Greenman, E., Bodovski, K., Reed, K.: Neighbourhood Characteristics, Parental Practices and Children's Math Achievements in Elementary School. The Pennsylvania State University, PA, United States (2011)
14. Gulati, S.: Technology-enhanced learning in developing nations: a review. In: The International Review of Research in Open and Distance Learning, vol. 9, No. 1(1), Feb. 2008, ISSN 1492–3831. http://www.irrodl.org/index.php/irrodl/article/view/477/1012 (2008)
15. Heeks, R., Molla, A.: Impact assessment of ICT for development projects: a compendium of approaches. In: Development Informatics, Working Paper 36 (2009)
16. Jonassen, D.H., Peck, K.L., Wilson, B.G.: Learning with Technology: A Constructivist Perspective. Merril (1999)
17. Jones, A., Naylor, R.: The Quantitative Impact of Armed Conflict on Education in Nigeria: Counting the Human and Financial Costs. CfBT Education Trust, Reading (2014)
18. Jonker, V., Wijers, M.: Th!nklets for mathematics education. Re-using computer games characteristics in educational software. In: Paper presented at the International Conference of the Learning Sciences (ICLS). From http://www.fi.uu.nl/en/icls2008/550/paper550.pdf (2008)
19. Kallaway, P.: The need for attention to the issue of rural education. Int. J. Educ. Dev. **21** (2001), 21–32 (2001)

20. Karweit, N.: Time-on-Task Reconsidered: Synthesis of Research on Time and Learning. Educational leadership May 1984, pp. 32–35 (1984)
21. Kegel, C.A.T., Bus, A.G., van IJzendoorn, M.H.: Differential susceptibility in early literacy instruction through computer game: the role of the dopamine D4 receptor gene (DRD4). Mind Brain Educ. **5**, 71–78 (2011)
22. Meijer, J., Karssen, M.: Effecten van het oefenen met rekentuin. Technisch rapport (Effects of practicing with Rekentuin. Technical report). In: Rapport 925, ISBN 90-6813-983-9. Amsterdam: Kohnstamm Instituut (2014)
23. Milo, B.: Mathematics Instruction for Special-Needs Students. Thesis University of Leiden/Heerenveen, Brouwer/Wielsma (2003)
24. Power, T., Gater, R., Grant, C., Winters, N.: Educational Technology Topic Guide. HEART Topic Guides. London: The Health & Education Advice & Resource Team (HEART). Department for International Development (DFID) (2014)
25. Reubens, A.: Early Grade Mathematics Assessment (EGMA): A Conceptual Framework Based on Mathematics Skills Development in Children—EdData II Technical and Managerial Assistance, Task Number 2 Contract Number EHC-E-02-04-00004-00 Strategic Objective 3. 31 Dec 2009 (2009)
26. Romiszowksi, A.J.: How's the e-Learning baby? Factors leading to success or failure of an educational technology. Innovation Educ. Technol. **44**(1), 5–27 (2004)
27. Saine, N.L., Lerkkanen, M.-J., Ahonen, T., Tolvanen, A., Lyytinen, H.: Computer-assisted remedial reading interventions for school beginners at risk for reading disability. Child Dev. **82**, 1013–1028 (2011)
28. Sarama, J., Clements, D.H.: Building Blocks for Young Children's Mathematical Development. Baywood Publishing Co., Inc, New York (2002)
29. Sriprakash, A.: Child-centered education and the promise of democratic learning: pedagogic messages in rural Indian primary schools. Int. J. Educ. Dev. **30**, 297–304 (2010)
30. The World Bank.: The Status of the Education Sector in Sudan. African Human Development Series, 66608 (2012)
31. Timmermans, R.: Addition and Subtraction Strategies, Assessment and Instruction. Radboud University, Nijmegen (2005)
32. UNESCO.: Toward Universal Learning What Every Child Should Learn (Report No. 1 of 3). Quebec, Canada (2013)
33. UNICEF.: Open and Distance Learning for Basic Education in South Asia (2009)
34. UNICEF.: Education Under Fire: How Conflict in the Middle East is Depriving Children of Their Schooling. Amman, Jordan (2015)
35. United Nations.: Measuring the Impacts of Information and Communication Technology for Development UNCTAD/DTL/STICT/2011/1 (2011)
36. Unwin, T.: ICT4D Implementation: Policies and Partnerships. In: Unwin, T. (ed.) ICT4D. Cambridge University Press, Cambridge, pp. 125–176 (2009)
37. Vogel, J., Vogel, D.S., Cannon-Bowers, J., Bowers, C.A., Muse, K., Wright, M.: Computer gaming and interactive simulations for learning: a meta-analysis. J. Educ. Comput. Res. **34**, 229–243 (2006)
38. Wagner, D., Day, B., Sun, J.: Recommendations for a Pro-Poor ICT4D Non-Formal Education Policy. Final report for Imfundo, DFID (2004)
39. Wagner, D., Day, B., James, T., Kozma, R.B., Miller, J., Unwin, T.: Monitoring and Evaluation of ICT in Education Projects—A Handbook for Developing Countries. *Info*Dev (Information for Development Program): www.infoDev.org (2005)

Empowering Vocational Math Teachers by Using Digital Learning Material (DLM) with Workplace Assignments

Diana Zwart, Johannes E.H. Van Luit and Sui Lin Goei

Abstract Digital Learning Material (DLM) are fast becoming a key instrument in teaching. The focus on E-learning systems with Digital Learning Material (DLM) is mostly on the medium and resources, instead of on the role of teachers. But like students, teachers also need to professionalize their digital competencies. And rather than describing how the teacher should take up his/her online role, teachers can also be trained by using online materials and experiencing DLM themselves. Therefore the first objective of this study is to design an e-learning system with DLM to train seven vocational math teachers. The second objective is to investigate the aspects of the learning activities in terms of "learning engagement," "teachers' self learning process" and "professional learning to enhance teachers' knowledge." In this research, teachers participating in this project had a 6 week online training in an e-learning system with DLM. After this, teachers completed the questionnaires on these aspects. Teachers were satisfied with the components "learning engagement" and "teachers' self learning process." They specifically appreciated the domain-specific literature and the online moderation of the expert teacher. With regard to the component "professional learning," teachers did not collaborate and only three teachers abstract and detach information from its original context and applied to it in new contexts by their questions and contributions in the forum discussions. For other teachers the "space to act and arrange" seemed too broadly. Future study should pay more attention to the development of assignments that require specific tasks in collaboration between the teachers.

D. Zwart (✉) · S.L. Goei
Centre for Human Movement and Education, Windesheim University
of Applied Sciences, Zwolle, The Netherlands
e-mail: dp.zwart@windesheim.nl

J.E.H. Van Luit
Department of Special Needs Education,
Utrecht University, Utrecht, The Netherlands

S.L. Goei
Faculty of the Behavioural and Movement Sciences, Vrije Universiteit Amsterdam,
Amsterdam, The Netherlands

Keywords Digital Learning Material (DLM) · Workplace learning · Collaborative learning · Mathematics · Vocational education

1 Introduction

With the rapid growth of online and blended approaches to teaching and learning, e-learning systems with Digital Learning Material (DLM) are fast becoming a key instrument in teaching. E-learning systems can be used to create environments with DLM in which tasks can be combined with whole-task practice that confronts learners with all or almost all of the constituent skills important for real-life task performance, together with their associated knowledge and attitudes [36].

With DLM, learners can discuss their ideas and conceptions from different perspectives in order to reconstruct and co-construct (new) knowledge that can be used for solving authentic and also complex problems, e.g., [28]. The tasks can be presented with words and pictures through combining visual and auditory presentation methods, which is called the multimedia principle [21]. This results in the 'modality effect', which means that more memory capacity is available when dual modalities—visual and auditory—are used, leading to better performance than using either the visually or auditory presentation alone [33]. Together with the opportunities to individual training and social settings [28] teaching with DLM provides better opportunities for training courses in education.

The role of teachers in DLM is still underexposed. The focus of e-learning systems with DLM in education is mostly on the medium and resources themselves. Chen [6] states that e-learning developers, academicians, and practitioners should strive to make the e-learning system with DLM meaningfully infused with educational principles so that it has real educational capabilities to fulfill a given unique learning expectancy. Working with DLM requires other teachers' skills. Instead of describing how teachers should take up their online roles, it might be better to train teachers using online materials, by experiencing the e-learning system with DLM themselves [29].

Despite insights into pedagogically appropriate uses, educational technology for teachers in everyday school settings are severely limited [22]. Only a very few studies investigated the impact of online training on teachers' knowledge. While training and experience increase competence and enhance teachers' beliefs in performance [7], the opportunities of training with DLM are that they can even train the digital competencies in addition to professional competencies. Therefore an objective of this study is to design an e-learning system with DLM to train teachers.

Math teachers in vocational education need to professionalize themselves, because mathematics has become a required part of the curriculum due to political decision-making; the math skills of the students in vocational education in the field were too low. Using the opportunities of DLM, a question arises from the design aspects what contributes to teachers' learning. Therefore, more specifically the objectives of this study are to design an e-learning system with DLM to provide vocational math teachers hands-on experiences, and to examine how it facilitates teachers' knowledge in vocational education.

1.1   Context: Professional Learning Through E-learning

In the Netherlands, vocational education is part of the formal education system, consisting of prevocational education for youngsters aged between 14 and 16 years, and upper secondary vocational education for youngsters aged between 16 and 20 years [3].

In vocational education, learning addresses concrete professional tasks, taking place in the workplace or inside vocational schools [3]. Vocational education teachers' competencies are therefore mostly workplace related. However, in 2010, the Dutch government introduced a reference framework[1] for mathematics in vocational education due to the insufficient math competencies of Dutch students [25]. Mathematics became an important part of the curriculum. So all of a sudden, vocational education teachers needed equipment for teaching mathematics because they did not always have sufficient numeracy skills and knowledge of mathematical didactics [26].

A critical aspect of learning to be an effective mathematics teacher for diverse learners is developing knowledge, dispositions, and practices that support building on student's mathematical thinking [35]. As teaching is a social process, discussing, collaborating, and interaction are also of great importance [17]. Collaboration is assumed to create a learning culture that helps to build a community in which further learning is supported and stimulated [19]. A way to structure collaborative argumentation in online learning is to provide learners with collaboration scripts which are defined as interventions that specify, distribute, and sequence learning activities, e.g., [18, 28]. Though collaboration scripts can be tailored, they do not necessarily support domain-specific knowledge acquisition [18]. This stresses the need for the disposal of domain-specific knowledge in DLM, specifically for mathematics so that teachers own prior knowledge and dispositions can be placed under critical scrutiny.

DLM should engage vocational education teachers in experiences based upon access to students that brings together imperatives of work, workplace, and learning [5]. A great deal of previous research into teaching in vocational education has been focused on learning through practice. DLM should therefore invite teachers by learning to practice tasks. By doing and experimenting, teachers not only gain new experiences but also apply new ideas, to put effort in improving their own professional practices.

---

[1]The reference framework consists of three fundamental-levels (F-level) and three target-levels: 1F, 2F, en 3F. The levels are guidelines which describe what students should know and be able to use during their specific school career.

1.2 Design Aspects for DLM in Vocational Math Teacher Training

Workplaces provide activities and interactions which are authentic in terms of the knowledge to be learnt for work that is undertaken in those settings [5]. Their social and physical settings offer contributions that are directly aligned with the activities to be undertaken. The more the learning activities are connected to the world of experience, the higher the intrinsic motivation and expectations are about teachers' own abilities [37].

Teachers can develop competencies through experience of engaging in practice, e.g., [23, 37]. In a study of [23] on engagement of teachers in learning activities, five general categories of learning activities were identified, (1) doing, (2) experimentation, (3) reflection on experiences, (4) learning from others without interaction, and (5) learning from others with interaction. Similarly [19], concludes that teachers themselves are key actors in directing and arranging their own learning processes and that they have to fulfill a new role by creating stimulating learning environments and by acting as facilitators in students' learning processes. But she makes a call for collaboration because feedback, new information, or ideas do not only spring from individual learning but to a large extent from dialogue and interaction with other people. Hargreaves [15] also found that professional learning occurs when (1) teachers collaborate with others, (2) when teachers learn things they are interested in, (3) when learning is connected to their school environment, and (4) when there is such thing as long term commitment to learning. In summary, one may suppose that teacher training includes learning activities that consist of performing, experiencing, reflecting, collaborating, and it should be connected to teachers' school environment and teachers own interest. Thereby the training should be process-based and it should come across with a world of experience. The described categories of the learning activities have limited utility with respect to teachers' skills to coordinate and reinforce the face-to-face and online components of learning [38]. They argue that teachers need to learn skills and require technological capabilities to implement flexible teaching and that learners want their personalized learning not only in the form of online activities. Face-to-face interaction in the tutorials and interactive and collaborative learning activities motivate them.

Motivation is a function of a person's acknowledgement on the task being performed, on the impact of the completion of the task, and one's ability to solve the task [9]. Similarly it is also well known that learners with higher self-efficacy are more motivated and successful on a task. Expectations that learners have to complete a mission successfully and the value of how they attach them is an important aspect of motivation and contribute to learning performance [10, 34]. Self-efficacy is, according to [7] a bridge between attitudes and competencies. This is critical because learners have to believe they can perform. This requires both the training and belief in oneself. The behavior of the teacher provides an important contribution to the development of the learners and their self-efficacy [27]. Nowadays teaching and learning are moving from teacher-centered pedagogies and practices to personalized learning in which learners are more actively involved in the learning process and where learners demand more flexibility through online and blended university courses [38].

1.3 Transfer of Knowledge and Skills for Promoting Teachers' Competence

Knowledge will not transfer to other contexts when it is wholly tied to the context of its acquisition [4]. Transfer leads, for example, to questions about the correctness of the school curriculum, that is, whether it includes the proper equipment for students to work, and to questions about whether learners can recall and apply what they learned at school when they are at work [2]. Without assuming extreme contextual dependence, one could still claim that there is relatively little transfer beyond nearly identical tasks to different physical contexts [1]. Representation and degree of practice are critical for determining the transfer from one task to another. Abstract instruction can therefore be ineffective if what is taught in the classroom is not what is required on the job. Often this is an indictment of the design of the classroom instruction rather than of the idea of abstract instruction in itself [1] though.

It is a crucial issue to find out when and under which conditions transfer problems occur [16]. This work describes a classification of four learning types which provide a basis for a new understanding of the transfer problems. It starts with cumulative learning (1) which is a form of learning characterized by being isolated formation, something new that is not a part of anything else. Then the most common form of learning is called assimilative learning (2) which means that the new element is linked as an addition to an established scheme. Accommodative learning (3) implies that one breaks down the established scheme and reconstructs in a way that allows the new situation to be linked in. Finally it describes transformative learning (4) which implies personality changes or changes of the organization of the self.

1.4 Knowledge Integration by Reflection

In their analogy to theories on transfer [3], classified three types of integration processes to take place during the performance of different professional learning tasks: (1) low-road integration, which is hypothesized to occur as a result of practice toward automatic performance, (2) high-road integration which means that learners abstract and detach information from its original context and applies to it in new contexts, and (3) transformative integration which is the connection between knowledge and skills occurs by means of critical (self) reflection. This not only refers to Illeris' accommodative learning [16] but also approaches Schön's reflection-in-action [31] suggesting that practitioners reflect during experience and make changes during action. Eraut [13] has challenged Schön's point of view on the grounds that it effectively ignores the important variable "time." "When time is extremely short, decisions have to be rapid and the scope for reflection is extremely limited. In these circumstances, reflection is best seen as a metacognitive process in which the practitioner is altered to a problem, continuing alertness (p. 145)."

Teachers often learn by critical individual reflection and by involving colleagues in particular challenging or problematic situations [24]. Eraut [13] describes that reflection is triggered by the recognition that in some respects the situation is not normal and therefore in need of special attention. This is adjacent to Engeströms' expansive learning [11, 12] which describes the learning when learners are involved in constructing and implementing a radically new, wider, and more complex object and concept for their activity. In essence, there is a contradictory pressure here. Teachers need time to reflect critically on their professionalization. The expansive learning is emerging today: the flexible teaching and learning provides more experiences a teacher can handle and reflect on. There is limited institutional and technological support to assist teachers in guiding and reflecting on the experiences originating from the flexible learning activities and the added time is not recognized or acknowledged, e.g., [38].

2 Exemplifying Design Aspect

2.1 E-learning and Teaching Skills

The development of the internet facilitates e-learning in the form of applications and reduces the boundaries to learning and compliments traditional teaching methods [20]. E-learning is superior to traditional approaches in its capability to update, store, retrieve, and share learning information. In an online community teachers can tell and retell 'stories' to develop 'collective knowledge' and forge a group identity [14].

Because vocational math teachers do feel to be not equipped for teaching mathematics and do not always have sufficient numeracy skills and knowledge of mathematical didactics, an online training in an e-learning system with DLM was designed. Therefore, this article exemplifies the case of a group of math teachers in vocational education who attend an online training to investigate whether the design aspects for e-learning activities enhanced their knowledge and how the teachers appreciated the aspects of the e-learning activities (see Fig. 1).

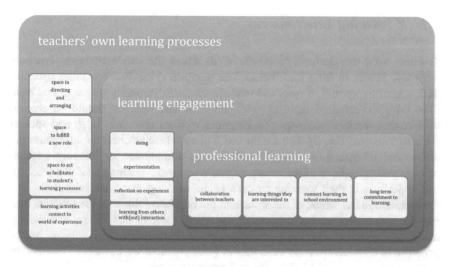

Fig. 1 Design aspects for e-learning activities in teachers' training

3 Context and Participants

The online training was designed with in an e-learning system with DLM in the platform N@tschool, at Windesheim University of Applied Sciences in Zwolle. There were seven teachers of a vocational education school in the Northern part of the Netherlands who participated in the online training for mathematics teacher training. The mean age of the participants was 49.2 years ($SD = 12.2$). The minimum age was 32 year and the maximum age was 61 year. The mean years of experience was 25.8 years ($SD = 14.1$) and the mean years of teaching math education was 5.8 years ($SD = 4.4$). Three teachers were male and four teachers were female. They came from different departments but all were assigned to teach mathematics in vocational education. The teachers, who had to professionalize themselves, were enrolled in a 6 week online training for the specific domain content and pedagogical content knowledge for vocational mathematics level 2F[1], the level of citizenship. Teachers need to professionalize themselves because mathematics is a required and new part of the curriculum. The goal was to gain (new) insights about pedagogics and didactics in teaching math practice by reading literature, discussion in the forum and performing assignments which consisted of applying the learned knowledge in practice tasks (integrated knowledge). The online training was introduced during a live class session, face-to-face. The teachers were online guided by a teacher who was a subject matter expert in guiding online training and mathematics.

3.1 E-learning Environment

Teachers used the platform 'N@tschool' to attend the online training. Teachers were able to log in from their home. The platform comprised the same structure each week: silverpoints, forum, and working space. Silverpoints are pages that were filled with the required features, with the aims that teachers should be able to achieve that week, domain-specific literature with internet links in it, and spoken presentations (see Fig. 2). The forum was a place where teachers could discuss the assignment of that week. Every week had an assignment for learning through practice and comprised of the theory of that week and a learning task that teachers had to perform in their working context. In the working space, teachers could find the literature in the article of the week and the portfolio folders of the teachers (see Fig. 3). The expert teacher moderated the online training three times a week, by using references to literature, domain-specific knowledge, and subject matter expertise in discussions. Also examples of good practices were shared. Teachers, as peers, were also allowed to answer or discuss questions.

3.1.1 Silverpoints

The silverpoints comprised of pages with written domain-specific knowledge about mathematics (pedagogical) content comprised with internet links, pictures, and spoken presentations (see Figs. 2 and 3). Because of that, teachers had the possibility to go more deeply into the materials (which was no obligation).

Fig. 2 Silverpoint with slides and spoken clips

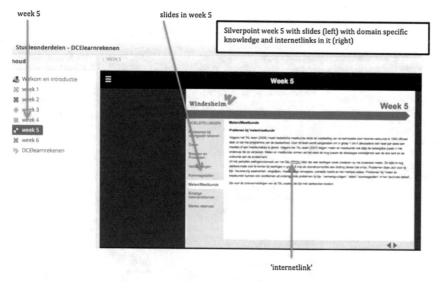

Fig. 3 Silverpoint with slides and 'internet links'

3.1.2 Forum Discussions with Assignments

The aim of the forum discussions was to discuss the assignment of the week (learning activities with practice tasks) and the theory of the week (see Figs. 4 and 5). Furthermore teachers could discuss the problems they bumped into during solving the assignment in practice. An example of an assignment is "analyse the test

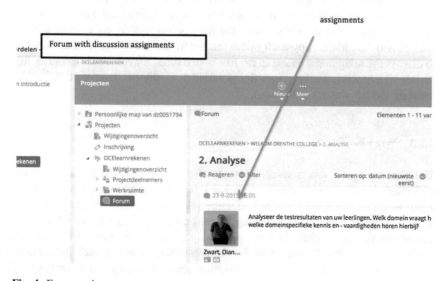

Fig. 4 Forum assignments

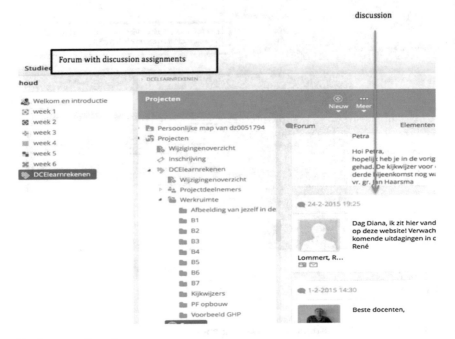

Fig. 5 Forum discussion

results of your students. In which domain do you find the most positive and negative remarkable results" and "what domain-specific knowledge and skills need attention?" And also, "how can you teach without the method to scaffold in the zone of the next development?" or "observe a lesson from a colleague or ask a colleague to observe your lesson. Then discuss the domain-specific content and the didactics used. What does this mean for the choices within the domain-specific content and the didactic approach of the teacher for the next lesson?" Teachers had to discuss the assignment online in the forum two times a week. And of course, like in the silverpoints, teachers had the possibility to discuss their own (practice) interests (which was no obligation).

3.1.3 Working Space

The working space was a space where teachers could find the literature on the addressed domain-specific content that week (Fig. 6). Every week teachers made assignments (learning activities with practice tasks) and uploaded their assignments in their portfolio folder in the working place. Furthermore, like in the silverpoints and in the forum, teachers had the possibility to upload interesting background information (see Fig. 7) related to the domain-specific content, such as opinions from newspapers on the Dutch national mathematics test (which was no obligation).

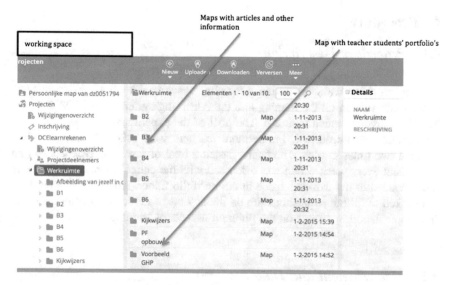

Fig. 6 Working space structure

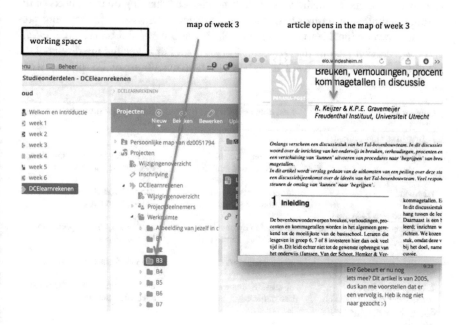

Fig. 7 Working space week 3

4 Procedure

4.1 Procedure Description

Prior to working with the online e-learning system with DLM, teachers were handed information on the training and the e-learning system with DLM during a live class session, which took 2 h. The goal of the experiment was explained and the content was introduced. Furthermore teachers were asked to discuss online minimal two times a week. The course lasted a total of 6 weeks. Teachers had an assignment every week which was discussed with the online teacher and the peers and uploaded in the working space in the portfolio folder. After 6 weeks teachers were asked to fill out a questionnaire on the aspects of the learning categories and their appreciation in ratings on the different aspects of learning.

4.2 Measurement and Data Analysis

The goal of the questionnaire was to analyze teachers' learning on the aspects of the learning activities in terms of 'learning engagement,' 'teachers own learning process' and 'professional learning' (see Fig. 8). It was a five-point Likert scale; *totally agree, agree, don't know, disagree, totally disagree*. For analyzing the results of the questionnaire the points were calculated with a maximum of 25 points, when

Categories	Learning activities	Items in questionnaire e.g. [32]
1. Learning engagement		
1.1 Doing	silverpoints, forum, discussions and café line	fascinating, inspiring, stimulant, enthusiastic, curious
1.2 Social	forum, discussions café line contribute two times a week	personal, social, fear time-consuming
2. Professional Learning		
2.1 Collaboration with peers	connection by own interests in the café line, forum and discussion line support colleagues	connection, contact peers and learn from peers support, feedback, positive
2.2 Expert teachers' competence communication	discussion online and assignments	clear, communication, personal, adequate, support
3. Teachers own learning process		
3.1 Forum discussions with teacher	discussion online and assignments	domain specific knowledge, stimulating
3.2 Flexible	silverpoints with links, working space and forum	personalized, outside trails, learning possibilities

Fig. 8 Categories and learning activities

aspects	literature (articles)	theory in silverpoints	structure (silverpoints)	internet links	working space	discussion (peers)	moderation (teacher)	assignment (forum)	Module as a whole
Rating (1-10)									

Fig. 9 Ratings of aspects of e-learning

teachers were satisfied and a minimum of five points, when teachers were not satisfied. Teachers also rated the e-learning aspects with a number from 1 (minimum, *not satisfied*) to 10 (maximum, *very satisfied*) (see Fig. 9). Five teachers completed the questionnaire. Two teachers, (out of seven) did not complete the questionnaire. The questionnaire stems partially from the questionnaire of [32].

5 Results

Teachers were overall satisfied with the aspects on "learning engagement" in the e-learning system with DLM. They appreciated the learning activities in "doing" as fascinating and curious. One teacher was not inspired. Four teachers were enthusiastic and stimulated. One did choose "neutral." The teachers appreciated it as personally and although the teachers felt very safe in the E-environment, three teachers did not value the e-learning system with DLM as very social, and two teachers choose "neutral." The teachers did not often interact with each other. So, not surprisingly, teachers pointed out that they were not satisfied about learning with and from their peers. Two teachers find the online training time consuming. *"In practice, it is quite difficult to contribute actively."*

Within the aspects on "professional learning" in the e-learning system with DLM, collaboration between peers was the least appreciated by teachers. Three teachers were "neutral" about the connection with their peers in the e-learning system with DLM and two teachers were not satisfied. They did appreciate the support of the peers but the feedback of the peers was not appreciated as high; they did not learn from their peers. *"I tried to contribute so it yield knowledge for every peer, but not all the peers did react or contribute. That's a pity because we all are busy and we did get professional hours to attend this online training"*. The teachers were very satisfied with the communication of the expert teacher, they found it clear, personal, and adequate. Most teachers were stimulated by the expert teacher and appreciated the support. *"Thank you very much for your support during the online training."* One teacher marked "don't know" here.

Teachers were satisfied about "self learning." The flexibility in the e-learning system with DLM, to walk outside trials, was appreciated: Teachers liked the possibility to upload interesting background information in the working space, to discuss their own (practice) interests in the forum and to go more deeply into the materials and domain-specific knowledge in the silverpoints. Teachers did learn and appreciated the learning possibilities. "*I liked to discuss the literature in the articles, it certainly gave direction to the way I think about math education now.*"

Teachers rated the different learning activities (in mean) as follows; literature 6.8, theory 7.0, structure of the content 6.6, internet links 7.4, working space 6.6, discussion with peers 5.8, moderation expert teacher 7.2, assignments 7.0 and the module as a whole 7.0. "*I liked to read the discussions but the environment was not very user-friendly. It required lots of clicks.*"

6 Conclusion

The aim of this chapter was to design an online training in an e-learning system with DLM to understand how it facilitates math teachers' learning. We used the case of a group math teachers in vocational education to exemplify the evaluation on the design aspects of an e-learning system with DLM described in terms of "learning engagement," "professional learning" and "teachers own learning process."

Overall we can conclude that teachers seemed satisfied about the design of the online training. They were satisfied about the components "learning engagement" and "teachers own learning process." They specifically appreciated the domain-specific literature and the online moderation of the expert teacher. This is also visible at the ratings on the aspects of the e-learning activities. The articles, written domain-specific theory, and internet links in the silverpoints, gained higher mean scores, than the other aspects. Although the moderation of the online expert teacher was also an important contribution to the satisfaction on the online training, this was in contrast to [19] who explains that teachers themselves are key actors in directing and arranging their own learning processes. However in this case, not all teachers acted as key actors, they did not collaborate much. Specifically connection and mutual contact was low. Wanner and Palmer [38] point out that "flexible learners," as self-directed and self-regulated learners, are not a given as not every learner is ready for, and open to, more personalized and self-regulated learning and to the shift from passive, to active, collaborative learning. This also counts for the teachers in this online training. They were passive and did not collaborate mutually.

One teacher was not satisfied with the professional learning component, but she did not contribute in the discussions at all. That is a pity because teachers can tell and retell 'stories' to develop "collective knowledge" and forge a group identity in an online community [14]. The question is whether this teacher developed competence through experience of engaging in practice as [23, 37] state. But according to [23] it is also possible that the teacher learned through "learning without interaction," because this teacher stated that she did learn and that she appreciated the literature in the articles.

Three teachers directed and arranged their own learning process. These teachers, as key actors, collaborated and arranged feedback, new information, and ideas [19].

They appreciated the online training as whole more positive. One teacher could not find his way in the e-learning system because he missed the first face-to-face meeting. He had to deal with the technology-related issues which includes teachers' abilities to use software and hardware [8]. *"Because I missed the first face-to-face meeting, I had to sort out many things by myself, that was frustrating."* By starting a face-to-face meeting the other teachers had to deal less with the technological issues and technical skills might be prerequisite for the proficient teacher level, as was visible in this research. All teachers pointed out that they had learned. For only two teachers time for studying was an issue. The studying in addition to providing education cost too much time. Sometimes the flexible teaching and learning provides more experiences a teacher can handle and reflect on. Though in this study there was institutional and technological support, the added time was not recognized or acknowledged, e.g., [38].

6.1 Contribution

The teachers appreciated the design aspects. One teacher was not active in discussions in the forum, but she did read the articles and also the theory in silverpoints. Though, according to [15, 19] learning appears by collaborating, it can therefore be assumed that this teacher learned with the help of the online training without interacting or collaborating.

This online training did not use a reflection tool which could contribute to learning as [24] state. Teachers often learn by critical individual reflection and by involving colleagues in particular challenging or problematic situations. According to the integration levels mentioned in the study of [3] this online training tends to a low-road integration level because knowledge and skills are conceptualized as a process in which knowledge and skills are increasingly connected by means of practice toward automatic performance. Knowledge and skills become increasingly implicit as they are internalized by learners. In addition, three teachers required reflection on their action [31] in the forum discussion. Those teachers were able to abstract and detach information from its original context and apply it in new contexts. Their connection between their knowledge and skills made them think consciously about what they were doing. So we might state that this online training offered a level of high-road integration, the second type of integration process according to [3] these three teachers. They did abstract and detach information from its original context and applied to it in new contexts by their questions and contributions in the forum discussion. For some teachers however, the "space to act and arrange" seemed too broadly. Perhaps these teachers were just "doing an activity simply for the enjoyment of the activity itself, rather than its instrumental value" [30] (p. 60).

6.2 Limitations and Future Improvement

The scope of this study was limited in terms of the development of learning. The only control visible was the logs into the system but that does not tell us much about the learning of teachers. Thereby the e-learning system with DLM only offered assignments that teachers could solve alone. It might be better to develop assignments which require collaboration between teachers that can be supported by tailored collaboration scripts which define interventions that specify, distribute, and sequence learning [18, 28]. Thereby, it might be that the expectations to the math teachers in vocational education were not clearly communicated in advance.

Future study on online teacher training in an e-learning system with DLM should pay more attention to the aspects of "professional learning," especially on the development of assignments that require collaboration between the teachers, and on exploring the ways in which reflection can contribute to the professional identity that enhance professional learning.

References

1. Anderson, J.R., Reder, L.M., Simon, H.A.: Situated learning and education. Educ. Res. **25**, 5–11 (1996)
2. Akkerman, S.F., Bakker, A.: Crossing boundaries between school and work during apprenticeships. Vocat. Learn. **5**, 153–173 (2012)
3. Baartman, L.K., De Bruijn, E.: Integrating knowledge, skills and attitudes: conceptualising learning processes towards vocational competence. Educ. Res. Rev. **6**, 125–134 (2011)
4. Berliner, D.C.: The near impossibility of testing for teacher quality. J. Teach. Educ. **56**, 205–213 (2005)
5. Billet, S.: http://www.unevoc.unesco.org/fileadmin/up/2013_epub_revisiting_global_trends_in_tvet_chapter4.pdf (2013)
6. Chen, J.L.: The effects of education compatibility and technological expectancy on e-learning acceptance. Comput. Educ. **57**, 1501–1511 (2011)
7. Christensen, R., Knezek, G.: Self-report measures and findings for information technology attitudes and competencies. In: Voogt, J., Knezek, G. (eds.) International Handbook of Information Technology in Primary and Secondary Education, pp. 349–365. Springer, New York (2008)
8. Compton, L.K.L.: Preparing language teachers to teach language online: a look at skills, roles, and responsibilities. Comput. Assist. Lang. Learn. **22**, 73–99 (2009)
9. Driscoll, M.P.: Psychology of Learning for Instruction. Pearson Education, Boston, MA (2005)
10. Eccles, J.S., Wigfield, A., Schiefele, U.: Motivation. In: Eisenberg, N. (ed.) Handbook of Child Psychology, pp. 1017–1095. Wiley, New York (1998)
11. Engeström, Y.: Learning by expanding. In: An Activity-Theoretical Approach by Developmental Research, Orienta-Konsultit, Helsinki (1987)
12. Engeström, Y., Sannino, A.: Studies of expansive learning: foundations, findings and future challenges. Educ. Res. Rev. **5**, 1–24 (2010)
13. Eraut, M.: Developing professional knowledge and competence. Routledge Falmer, London/New York (1994)
14. Gray, B.: Informal learning in an online community of practice. J. Distance Educ. **19**(1), 20–35 (2004)

15. Hargreaves, E.: The validity of collaborative assessment for learning. Assess. Educ. Princ. Policy Pract **14**(2), 185–199 (2007)
16. Illeris, K.: Transfer of learning in the learning society: how can the barriers between different learning spaces be surmounted, and how can the gap between learning inside and outside schools be bridged. Int. J. Lifelong Educ. **28**, 137–148 (2009)
17. Kahveci, M., Imamoglu, Y.: Interactive learning in mathematics education: Review of recent literature. J. Comput. Math. Sci. Teach. **26**, 137–153 (2007)
18. Kollar, I., Ufer, S., Reichersdorfer, E., Vogel, F., Fischer, F., Reiss, K.: Effects of collaboration scripts and heuristic worked examples on the acquisition of mathematical argumentation skills of teacher students with different levels of prior achievement. Learn. Instr. **19**, 149–170 (2014)
19. Kwakman, K.: Factors affecting teachers' participation in professional learning activities. Teach. Teach. Educ. **19**, 149–170 (2003)
20. Lee, L.T., Hung, J.C.: Effects of blended e-learning: a case study in higher education tax learning setting. Human-centric Comput. Inf. Sci. 5, 1–15 (2015)
21. Mayer, R.E.: Multimedia Learning. Cambridge University Press, New York (2001)
22. McKenney, S.: Designing and researching technology-enhanced learning for the zone of proximal implementation. Res. Learn. Technol. **21**, 1–9 (2013)
23. Meirink, J.A., Meijer, P.C., Verloop, N.: A closer look at teachers' individual learning in collaborative settings. Teach. Teach. Theory Pract. **13**, 145–164 (2007)
24. Meirink, J.A., Meijer, P.C., Verloop, N., Bergen, C.M.: How do teachers learn in the workplace? An examination of teacher learning activities. Eur. J. Teach. Educ. **32**, 209–224 (2009)
25. Ministry of Education, Culture and Science. https://www.rijksoverheid.nl/onderwerpen/taal-en-rekenen/inhoud/referentiekader-taal-en-rekenen (2010)
26. Ministry of Education, Culture and Science. https://www.rijksoverheid.nl/binaries/rijksoverheid/documenten/kamerstukken/2015/10/06/kamerbrief-over-invoering-van-referentieniveaus-taal-en-rekenen/kamerbrief-over-invoering-van-referentieniveaus-taal-en-rekenen.pdf (2015)
27. Montague, M., Rinaldi, C.: Classroom dynamics and children at risk: a follow up. Learn. Disabil. Q. **25**, 75–83 (2001)
28. Noroozi, O., Biemans, H.J.A., Weinberger, A., Mulder, M., Chizari, M.: Scripting for construction of a transactive memory system in multidisciplinary CSCL environments. Learn. Instr. **25**, 1–12 (2013)
29. Roscoe, R.D., Chi, M.T.H.: Tutoring learning: the role of explaining and responding to questions. Instr. Sci. **36**, 321–350 (2008)
30. Ryan, R.M., Deci, E.L.: Intrinsic and extrinsic motivations: Classic definitions and new directions. Educ. Psychol. **25**, 54–67 (2000)
31. Schön, D.A.: Educating the reflective practitioner. Jossey-Bass, San Francisco, CA (1987)
32. Smits, A.E.H.: Ontwerp en implementatie van de masteropleiding special educational needs via e-learning (Design and implementation of the master special educational needs through e-learning). Ipskamp, Enschede, The Netherlands (2012)
33. Sweller, J., Van Merriënboer, J.J.G., Paas, F.G.W.C.: Cognitive architecture and instructional design. Educ. Psychol. Rev. **10**, 251–296 (1998)
34. Thoonen, E.E.J., Sleegers, P.J.C., Peetsma, T.T.D., Oort, F.J.: Can teachers motivate students to learn. Educ. Stud. **37**, 345–360 (2011)
35. Turner, E.E., Drake, C., McDuffie, A.M., Aguirre, J., Bartell, T.G., Foote, M.Q.: Promoting equity in mathematics teacher preparation: a framework for advancing teacher learning of children's multiple mathematics knowledge bases. J. Math. Teacher Educ. **15**, 67–82 (2011)
36. Van Merriënboer, J.J.G., Kirschner, P.A.: Ten Steps to Complex Learning. Lawrence Erlbaum, New Jersey (2007)
37. Wang, M., Eccles, J.S.: Social support matters: Longitudinal effects of social support on three dimensions of school engagement from middle to high school. Child Dev. **83**, 877–895 (2012)
38. Wanner, T., Palmer, E.: Personalising learning: Exploring student and teacher perceptions about flexible learning and assessment in a flipped university course. Comput. Educ. **88**, 354–369 (2015)

The Odyssey Game

Jaap van der Molen, Henk Wildeman, Sui Lin Goei
and Alvin Sebastian Hoo

Abstract A case study presents an approach to the design of serious games through intercultural collaboration between students of two different institutions of higher education. Dutch students provided input regarding instruction about literature, and Singapore students shared knowledge of game design. The group produced multiple chapters of a game to be used in classrooms for literature instruction. The game focuses on *The Odyssey*, which is an important source text for modern literature. Then primary school, secondary school, and teacher education students played the game; research data were obtained to determine the target group for whom the game is suitable.

Keywords Serious games · Game-based learning · Literature education · Student exchange · International collaboration · Interdisciplinary design · E-learning · Game design · Educational design · Educational games · Game research

J. van der Molen (✉) · H. Wildeman · S.L. Goei
Department of Human Movement and Education, Windesheim University of Applied
Sciences, Zwolle, The Netherlands
e-mail: j.van.der.molen@windesheim.nl

S.L. Goei
e-mail: s.l.goei@vu.nl

S.L. Goei
Faculty of Behavioural and Movement Sciences, Vrije Universiteit Amsterdam, Amsterdam,
The Netherlands

A.S. Hoo
School of Interactive & Digital Media, Nanyang Polytechnic, Singapore, Singapore
e-mail: Alvin_Sebastian_Hoo@nyp.edu.sg

© Springer Science+Business Media Singapore 2017
Y. Cai et al. (eds.), *Simulation and Serious Games for Education*,
Gaming Media and Social Effects, DOI 10.1007/978-981-10-0861-0_7

1   Introduction

This chapter describes the collaborative development of a serious game for Dutch literature instruction (i.e. the Odyssey game) by students and staff of two institutes for higher education, in the Netherlands and in Singapore. Results of a pilot test of the game's usefulness in a Dutch educational setting are also presented.

The Dutch institute of higher education involved was Windesheim University of Applied Sciences, specifically the Department of Secondary Teacher Training for the Dutch language, together with the Department of Human Movement and Education. Windesheim University is located in Zwolle, a middle-sized town in the northeast of the Netherlands. Windesheim University focuses on high quality higher education, research and entrepreneurship. With over 20,000 students and 2000 employees in three locations (Windesheim, Zwolle and Almere), it is one of the largest universities of applied sciences in the Netherlands. The participating institute of higher education in Singapore was Nanyang Polytechnic (NYP), specifically the School of Interactive & Digital Media (SIDM). SIDM is recognised by foreign institutions and media professionals to be one of the finest schools for interactive and digital media studies in Asia. Each year, SIDM trains about four hundred aspiring students for the thriving digital media industry in the region. SIDM's game courses equip the students with a solid foundation in integrating game art and design with rapidly changing game technologies, across different platforms such as PCs, consoles, mobiles, and the like. NYP also manages a national incubation centre, the Games Solution Centre (GSC), a one-stop resource centre that provides a rapid prototyping development environment for Singapore-based small-medium game enterprises to develop their games.

This cross-cultural project was set up with the specific aim of bringing the best of both worlds together, in terms of validation of technology—in this case a serious game—in an educational setting, the design of new pedagogies for the use of technology in classroom situations, and last but not least, the making of the serious game itself by designing and programming it.

It is helpful for future teachers to experiment with modern technology in relation to the instruction of literature. Teacher educators and educational researchers at Windesheim University have the pedagogical and instructional know-how, but not the technological knowledge and means to build a serious game. The faculty at Nanyang Polytechnic are experts in developing these types of applications, but lack knowledge regarding the validation and deployment of a serious game in educational settings. For students at Nanyang Polytechnic, the value of this collaborative project was that they learned to develop a serious game that met the demands of a specific target group in a concrete educational situation, in this case, students in literature lessons. For the Dutch students, it was a valuable experience to explain their ideas and thoughts to collaborators from other disciplines. Both groups of students developed general skills such as project management, intercultural collaboration and communication.

The goal of this endeavour was to use mutual expertise and insights in the fields of game development, education and research. In the first half of 2013, both institutions started preparing for this cross-cultural project. Teacher educators at Windesheim University were asked to send in ideas for the content for this project, and in September 2013 the first group of students kicked off the Odyssey project with the aim of developing a serious game for literature. What follows is a brief explanation of the literary and educational concepts on which the game is based.

1.1 Literature Education

A modern text sometimes contains traces of an older text. This more or less complex relationship between texts is called *intertextuality*. In Dutch schools, literature education usually focuses on the strategy of *close reading*, where students learn to analyse poems or stories thoroughly without using information outside the text, such as biographical, historical or cultural background. Intertextual reading can be a valuable addition to this strategy, because it makes students aware of the fact that these literary texts are not isolated artistic expressions, but are part of a rich tradition. "Intertextuality is a crucial element not only in the attempt to understand literature in general but also in our attempt as educators to enhance our students' literary reading by locating it into a motivated and meaningful classroom context" [1, p. 180].

Intertextuality is a topic in the curriculum of teacher education for the subject of the Dutch language at Windesheim University of Applied Sciences in the Netherlands. This teacher education is preparing students to become teachers in secondary and vocational education. Students learn to analyse and interpret modern contemporary literary texts that refer to older texts such as the Bible or ancient myths. To do that correctly, they need to be familiar with these old stories [1].

An example of such an old and influential story is Homer's work, *The Odyssey*, a classic story that is often referred to in western prose and poetry [2]. Learning a new domain, such as becoming familiar with *The Odyssey*, requires the learner to take on a new identity, and a good video game captures players through identity [3]. Therefore a digital role-playing game was chosen: the gamer identifies with the hero and must solve all types of problems before he can make progress and complete the game. It could also be called a click-and-play game: one clicks on objects, persons, spaces and texts to make progress. In this way, the gamer is able to reconstruct the story actively. Thus, learners can familiarise themselves with *The Odyssey* in a simple and attractive way. The game provides an innovative learning experience for students: they do not begin by reading the story, but by identifying with a character and performing actions.

A total of five chapters retelling the story of *The Odyssey* were conceived and planned to be developed. To date (2015), all five chapters are nearly finished. The focus of this text is chapter one, because it was pilot-tested within several classrooms. It is an adaptation of an episode from Homer's *Odyssey*, namely when

Odysseus is in the cave of the giant Cyclops and his escape. The other chapters will be validated and evaluated in 2016.

1.2 Are Serious Games Serious Enough?

Games can be played for entertainment, but games are also linked to various forms of learning: skill acquisition, knowledge acquisition, affective and motivational outcomes and behaviour change outcomes [5]. The use of games for these kind of purposes has led to the name "serious games", games that are developed for purposes other than just entertainment.

Although games seem to be suitable for educational purposes, not much research has been done on how they are actually used in teaching [6]. In their literature review, Connolly et al. [5] conclude that it is difficult to relate learning outcomes directly to a game. Backlund and Hendrix [6] note that not much research has been done on how games are actually used in teaching. There seems to be some scepticism among students and teachers. Practical barriers include the difficulty of seeing how a game can be relevant with respect to the curriculum and of persuading stakeholders of the benefits of using games, the lack of time for teachers to learn how to use games as teaching tools, and technology barriers related to the school's IT equipment. When games are used, the teaching context and the pedagogical-instructional activities surrounding the game are important [6]. That means the effects of a game can only be measured when the educational context is taken into account.

1.3 Game-Based Learning

A game is a goal-directed and competitive activity within a framework of agreed rules that constantly provides feedback to enable players to monitor their progress towards the goal [7]. Games also "situate the meanings of words in terms of the actions, images, and dialogues that they relate to, and show how they vary across different actions, images, and dialogues" [3, p. 8]. Gee calls this the Text Principle: the story is not understood verbally, but in terms of embodied experiences that contribute to a better verbal understanding of the source text and related texts. The Intertextual Principle is also an underlying element in game play: after achieving embodied understandings of a text the learner understands that the text is part of a family of related texts [4]. To make this experience successful, the game must meet four criteria: it must be fascinating, efficient and easy to use, it must use technology that is available everywhere and it must concern everyday problems [8].

Gee's two principles were important guidelines during the development of the Odyssey game because they stress the relationship between games and language, which is of course an interesting approach for literature-related education. In accordance with the Text Principle, students identify with Odysseus and carry out actions on his behalf, which makes them familiar with the story without reading it. The Intertextual Principle is also important: together with different literary texts, the game can be used in an educational context to stimulate intertextual reading.

Mortara et al. [9] claim that a serious game seems to be particularly suited for the affective domain, in terms of identification with a character and understanding his feelings, problems and behaviours. The Odyssey game could therefore be used in different educational contexts, for example, literary education, cultural history lessons or lessons philosophy or ethics.

2 The Making of the Odyssey Game

Here a brief overview is given on the development of the first chapter of the Odyssey game, from September 2013 to June 2014.

2.1 Preparation

September 2013–December 2013

In September 2013, four student teachers from Windesheim University were selected to work on this project. Their goal was to develop a serious game about *The Odyssey*. To accomplish this, they needed to collaborate with three Game Development students from Nanyang Polytechnic. A project week in January 2014 was planned, to do some serious live project work. In preparation for this project week, both groups of students prepared by doing literature research and literature reading, and having project meetings with their supervisors.

As the new game would revolve around *The Odyssey*, the Dutch student teachers started reading Homer's *Odyssey*. Simultaneously, they started to learn about serious gaming. Because the project also focused on new pedagogies, they followed two MOOCs (Massive Open Online Courses), "Video Games and Learning" [10] and "Online Games: Literature, New Media, and Narrative" [11]. Students learned about Gee's principles and about intermediality, the way in which literature, film, and games engage in the basic human activity of storytelling. The principles of serious gaming proved their value immediately, because the students started selecting parts of *The Odyssey* that were suitable for using in the game. The MOOC about online games addressed *The Lord of the Rings* in different media, which served as an example, because there is also a role-playing game for this book trilogy.

After following the MOOCs and readings of *The Odyssey*, literature on serious gaming and intertextuality, the group was able to select a few episodes from *The Odyssey*. They were selected based on suitability for in-game use and evidence of their traces in modern texts. The students gathered poems containing traces of Odysseus' story, which helped them with this task. Meeting with their teacher and working together via Facebook and Skype, the students started organising their preparation for the project week. They made overviews of the selected episodes from *The Odyssey* that provided useful information related to their suitability for game development. This prepared them to meet with the students in Singapore to really start developing the game.

The story of *The Odyssey* was not included in the study of literature in Singapore. Furthermore, literature itself was somewhat foreign to the group of Singapore students, as none of them majored in that subject. The students started to prepare for the planned live work week with the Dutch student teachers by reading about and getting more resources on the story of *The Odyssey*. The students also started to create concepts and images for the game. They studied the character traits and temperament of the hero, Odysseus, and proposed various designs for the character.

2.2 Collaboration

In January 2014, a delegation of students and staff from Windesheim visited Singapore to work on the game. Both student groups were well prepared and highly motivated to produce a serious game. They communicated in English, which was also the language used in the initial version of the game itself. It quickly turned out that the initial idea of a fully playable game consisting of all selected episodes was too much work. It would never be ready in time. Therefore, the decision was made to make one playable chapter, consisting of one episode of *The Odyssey*. The game would focusses on Odysseus in the Cyclops's cave. Because of the preparation done back at home, the Windesheim students were able to quickly collaborate with the NYP students. After this chapter was completed, more chapters were created; a total of five chapters have been developed.

At NYP, the beginning of any game development process starts by coming up with a Game Design Document (GDD). Some call it the Game Bible or Design Bible, while the others call it Game Overview Document; for decades, the GDD has been used to provide a unified vision for game production to countless game developers and artists. It is a document with all of the possible steps in the game itself. The game programmers and designers use this to create the actual game. It can be seen as the construction manual for the game. The Windesheim students were able to put all of their knowledge into this GDD, while the NYP students continued to further develop concepts and game designs for the game chapters and

Fig. 1 Collaboration during the project week

other characters. Together, all seven students finished the first version of the GDD by the end of the project week (Fig. 1).

After the project week, the collaboration continued long-distance via video-conferencing, email and other digital communication. The groups worked on the next versions of the GDD by commenting on, and improving, the previous versions. The GDD also contained all of the text in the game, which was written by the Windesheim students. They could also make sure all the aspects of the story were right in the game. The NYP students learned to incorporate specific story details into the game. They also provided the overall structure, because the Windesheim students were less experienced at producing a game. By the end of May 2014, the first actual playable chapter was ready. It was developed on the basis of the con-stantly improving GDD, proving that the collaboration had been productive during the project week, as well as the weeks that followed (Fig. 2).

2.3 Other Activities

Every three months, a new batch of NYP students took over the game development from the current batch of students. Therefore, continuous communication with the Windesheim students remained vital for the success of the project. The NYP stu-dents continued to build the other chapters of the game with the continued feedback from Windesheim students and supervisors. The development of the first chapter of the game established a crucial framework and pipeline for the production process.

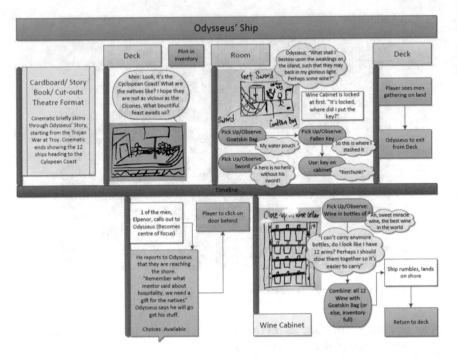

Fig. 2 Part of a game design document

New mini-games were designed for every new chapter, and the NYP students continued to push their capabilities to improve on the visuals of the game.

The game was originally designed and made for the PC; however, the decision was made to switch the target platform to mobile devices, due to the wide adoption of mobile devices as a teaching aid in schools these days. Such a drastic decision to change platforms would not have been possible in the past, but because the game was developed using an up-to-date game engine that allows flexibility in switching platforms, the students were able to deliver the game for mobile devices.

The game also supported play in two languages, English and Dutch. First, the NYP students designed and planned the English dialogues, which were written in a common text format and sent it over to the Windesheim students for translation. Once the translation was completed, it was sent back to the NYP students to integrate; after that, the game could be played immediately in the Dutch language as well.

Apart from these activities Windesheim students also played other important roles. They designed lessons in which they incorporated the game that had been developed. Furthermore, they presented their progress and outcomes to their fellow students, their teachers and others. Later on, they were able to assist with the research on use of the game by helping out in classrooms and processing test results. Not only did they help to build this serious game, but they also helped out during the entire project. The game is only useful in its proper context, and the students helped to establish that context as well.

2.4 Learning Outcomes

The students from Windesheim University learned a great deal from this project. Future teachers need to be future-proof. This project prepared the students involved for the future. They learned about a very promising new teaching method, serious gaming. And because they were involved in this project, they were able to experience first-hand how new types of instruction come to be. They learned how to do research on new pedagogical methods and actually build on this research, making educational material that is future-proof. Furthermore, they learned how to continue to develop themselves via MOOCs and other self-teaching methods. This is a key skill for teachers. A teacher needs to be able to think critically about himself and improve his methods. The students were able to grow in this respect.

Along with this, they learned about international collaboration. The world is becoming more and more connected, which means that we can use knowledge from all over the world. Being able to work with people from different cultures and learning from your differences is critical to succeeding in the connected world. Through this project, students were able to experience what it is like to work with colleagues from across the globe. This unique opportunity was an invaluable experience for the students. They learned how to communicate with people from totally different cultures, worked with them and were able to learn from their skills and know-how. This project delivered on a fully playable and usable educational game. Additionally the students learned a lot about instructional innovation, research and international collaboration.

3 Research

In parallel to its development, the Odyssey game was tested in various educational settings to find out the target group for whom the game is suitable.

3.1 Methodology

Different groups of students played the game in the period from August to November 2014: a group of 6th grade children grade in a Dutch primary school (11–12 years old), five Dutch students at an international school in Singapore (13–15 years old), 17 students in a Dutch secondary school (15–17 years old) and 20 Dutch teacher education students (21–23 years old). After playing the game in the classroom, most of the students (except those at the Dutch primary school) completed a questionnaire [12].

The goal of the research was to find out if the game is usable, meaning whether students find the game possible, attractive and meaningful. The survey (see the Appendix) includes eight questions about user experiences in general and 24 statements about five specific dimensions:

1. Intrinsic motivation (the extent to which a student experiences fun, pleasure and enjoyment)
2. Self-efficacy (the extent to which the gamer thinks he has enough skills to play the game)
3. Utility (value of the game for learning more about Odysseus)
4. Feeling of choice (the degree to which a student experiences actions as his own choice)
5. Engagement and concentration (the degree of focus and emotional involvement)

Students indicated whether they agreed or disagreed with different statements about the five dimensions.

3.2   Results

Students at the secondary school reported low *intrinsic motivation*, namely a mean of 1.6 on a scale from 1 to 4. The teacher education students had a mean score of 2.7 and the students at the international school showed the highest mean score, 3.7.

On *self-efficacy*, the extent to which the gamer thinks he has enough skills to play the game, the secondary school and teacher education students scored more or less neutral, with means of 1.7 and 2.2, respectively. All four students at the international school scored 4 for self-efficacy, which means they completely agreed with the positive statements about their game competences.

The third motivational construct was *utility*, formed by statements about the value and usefulness of the game for educational purposes. Students at the secondary school disagreed with these statements, with a mean score of 1.3. Students at the international school and student teachers were more positive about this, with mean scores of 2.8 and 2.6, respectively.

The motivational construct of *feeling of choice* refers to the degree to which the gamer feels he is free to choose his actions. The mean scores varied from totally disagree to neutral: a mean of 1.0 for students at the international school, 1.5 for those at the secondary school, and 2.3 for student teachers.

Finally, the students answered questions about *engagement and concentration*, the extent to which they could focus on playing the game and were involved in it. The secondary school students again had a low mean score, while the mean score for students at the international school was high and students of teacher education had a mean score between neutral and agree.

The mean scores for all five dimensions by school are visualised in Fig. 3:

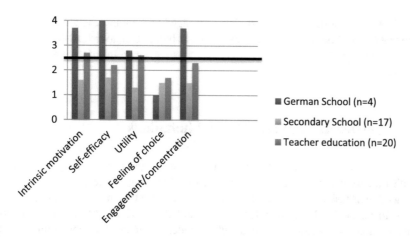

Fig. 3 Total score

A mean score of 2.5 or higher for each element is considered as positive and a lower score as negative. The Dutch primary school students did not complete the questionnaire, but the teacher inventoried their comments after playing the game in the classroom and shared his observations. The children were positive about the fact that you have to solve problems, search items and discover things and about the fact that you have to collaborate with others to complete the game. Some students judged that the game was too slow and some would like to have more clues. The teacher observed the students and noted that a few students gave up after a while, but that most students were determined to complete the game. In his own words: "I never saw a group of girls work so hard. There was a smell of sweat round the computers and they did not stop. That was wonderful to see."

At the Dutch secondary school, the game was used in lessons on cultural history. Students were told about the story of *The Odyssey* and were instructed to make a cultural artefact about this subject. The game functioned as an introduction. A student teacher assisted the teacher and observed the students during the game. The reactions of the students ranged from enthusiasm to little interest. In this setting, students also collaborated with each other: "It was fun to see that students actually helped each other, both enthusiastic and less interested students. Could someone not find a solution? Then someone else knew what had to be done. In this way they were very active." At that point, the game contained a few bugs that made it difficult to make progress in the game.

The teacher education students played the game in a computer lab. They collaborated intensively during the game and formed small groups on their own initiative. Only one student played the game alone and was, in her own words, not motivated. Because only a few students managed to complete the game, a group of students played the game on their own time at home. Collaboration then continued online through Facebook: they asked each other for help and gave each other directions.

4   Conclusions, Discussion and Limitations

Students who were involved in the development of the game were positive about both process and product, in the end. At the beginning of the process they experienced their task as open and vague, but after a while they took more initiative and they managed to shape their ideas more and more. There was a lot of creative energy that they could have channelled better, in hindsight. For similar future collaboration processes, communication between students should be given attention: clear arrangements about timing and means of communication are necessary to ensure smooth progress. What dominated was the feeling that this way of studying is "refreshing and attractive", as a Dutch student wrote in a reflection report.

The pilot research results show that all students were more or less motivated to play the game. The students at the international school were positive about engagement and concentration, the teacher education students were neutral on this point and the secondary school students were negative. With regard to the usefulness of the game, the students at the international school and the teacher education students were positive and the secondary school students disagreed. Only the students at the international school showed strong self-efficacy. Some students thought they were skilled enough, felt engaged and could concentrate. Students

Fig. 4 Screenshot of The Odyssee

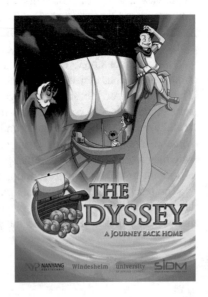

lacked a feeling of choice, which is no doubt a consequence of the fact that the story of *The Odyssey* is fixed (Fig. 4). Nevertheless, it is worth paying attention to this point in the further development of new game chapters.

The data provided a lot of information, but there were also a few limitations. First, the data reflected only the experiences of a relatively small group of students regarding only one episode of the game. Moreover, they did not play the game in a well thought out educational context. Games are more effective when they are incorporated within instructional programmes that include debriefing and feedback [13]. It is therefore also important to describe the educational context (preparation, subject, learning objectives, and so on), because these preconditions can influence the students' opinions and performance.

Acknowledgments We thank the dean, Harry Frantzen, of Windesheim University of Applied Sciences, Zwolle, the Netherlands, and the director, Daniel Tan, and deputy director, Albert Lim, of Nanyang Polytechnic, Singapore for making this project possible. Participants in this joint project were Sanne van den Bosch, Gea Pool, Rick Slaathuis and Henk Wildeman (students at Windesheim University of Applied Sciences, The Netherlands) and Bryan Koh Li Min, Desmond Poh Yongqiang, Noi Yuan Sheng, Xie Si'En, Lim Aik Sheng and Lee Choon Kwan (students at Nanyang Polytechnic, School of Interactive & Digital Media). Philippe Blanchet, Alvin Sebastian Hoo (Nanyang Polytechnic) and Wim Trooster, Sui Lin Goei and Jaap van der Molen (Windesheim University) were responsible for project management and research.

References

1. Kalogirou, T., Economopoulou, V.: Building bridges between texts: from intertextuality to intertextual reading and learning. Theoretical challenges and classroom resources. http://www.exedrajournal.com/exedrajournal/wpcontent/uploads/2013/01/14-numero-tematico-2012.pdf (2012). Retrieved on 24 Nov 2015
2. Ghesquire, R.: Literaire verbeelding. Een geschiedenis van de Europese literatuur en cultuur tot 1750. Leuven/Voorburg: Acco, (2005)
3. Gee, J.P.: Good video games and good learning. http://ocw.metu.edu.tr/pluginfile.php/2380/mod_resource/content/0/ceit706/week7/PaulGee.pdf (2005). Retrieved on 24 Nov 2015
4. Gee, J.P.: What Video Games Have to Teach us About Learning and Literacy. Palgrave MacMillan, New York (2007)
5. Connolly, T.M., Boyle, E.A., MacArthur, E., Hainey, T., Boyle, J.M.: A systematic review of empirical evidence on computer games and serious games. Comput. Educ. **59**, 661–686 (2011)
6. Backlund, P., Hendrix, M.: Educational games—Are they worth the effort? A literature survey of the effectiveness of serious games. In: 5th International Conference on Games and Virtual Worlds for Serious Applications. http://ieeexplore.ieee.org/Xplore/home.jsp (2013). Retrieved on 24 Nov 2015
7. Wouters, P., van der Spek, E., van Oostendorp, H.: Current practices in serious game research: a review from a learning outcomes perspective. http://www.cs.uu.nl/docs/vakken/b3elg/literatuur_files/Wouters.pdf (2009). Retrieved on 24 Nov 2015
8. Fullan, M.: Stratosphere: integrating technology, pedagogy and change knowledge. Pearson Canada Inc, Toronto (2013)
9. Mortara, M., Catalano, C.E., Belloti, F., Fiucci, G., Houry-Panchetti, M., Petridis, P.: Learning cultural heritage by serious games. J. Cult. Heritage **15**, 318–325 (2013)

10. University of Wisconsin–Madison.: Video and games and learning. https://www.coursera.org/course/videogameslearning (2013). Retrieved on 24 Nov 2015

11. Vanderbilt University Online Games: Literature, new media, and narrative. https://www.coursera.org/course/onlinegames (2013). Retrieved on 24 Nov 2015

12. Trooster, W.J., Goei, S.L., Ticheloven, A.H.L., Oprins, E., Visschedijk, G., Corbalan, G., van Schaik, M.: The effectiveness of LINGO Online. A serious game for English pronunciation. Internal report. Windesheim University of Applied Sciences, Zwolle (2015)

13. Hays, R.T.: The Effectiveness of Instructional Games: A Literature Review and Discussion. Naval Air Warfare Centre Training, Orlando (2005)

Social Development for Children with Autism Using Kinect Gesture Games: A Case Study in Suzhou Industrial Park Renai School

Zengguo Ge and Li Fan

Abstract Suzhou Industrial Park Renai School has taken the lead in the practice of autism education by using Kinect Gesture Game in the mainland China. This study aims to investigate the benefits of social development for children with autism by applying the innovative gesture-based games focusing on human–computer interaction, competitive cooperation, and emotional experience. Initial results show that Kinect Gesture games have the potential to help children with autism in terms of the development of verbal skills, communication skills, peer relationship, as well as interests.

Keywords Children with autism · Educational practice · Kinect gesture game · Social development

1 Introduction

Since 2002, the Horizon Report by the New Media Consortium has become an important source world widely on information technology developments for education. This annual report predicted and described a number of new technologies which may play a critical role in education around the world over the next 5 years. From the Horizon Reports (K-12 Edition) in recent years, areas with keywords such as 3D technology and Gesture Games have been in hot spot.

Educational practice of using Gesture Games in teaching and learning has been recognized as a new way to meet needs of children with disability worldwide [1, 2]. In recent years, a number of theoretical and practical explorations have been conducted in mainland China. Lin et al. [3] investigated the platform of Gesture Game in sensory integrated training, Cao et al. [4] looked into medical rehabilitation. Ma [5] studied parent–child interaction, and Liu and Shi [6] showed

Z. Ge · L. Fan (✉)
Suzhou Industrial Park Renai School, Suzhou Industrial Park,
211 Yangdong Road, Suzhou 215021, Jiangsu, People's Republic of China
e-mail: sipras@163.com

© Springer Science+Business Media Singapore 2017
Y. Cai et al. (eds.), *Simulation and Serious Games for Education*,
Gaming Media and Social Effects, DOI 10.1007/978-981-10-0861-0_8

113

performance enhancement using Gesture Games. In the year of 2013, Kinect Gesture Games were first introduced to classroom teaching in Suzhou Industrial Park Renai School. We have attempted to apply this new technology for autism education with positive results so far.

This chapter describes the educational practice of using Kinect Gesture Games in a special school. By playing with this new educational product, children with autism could improve their social developments, verbal skills, communication skills, peer relationship, as well as positive interest and behavior.

1.1 Gesture Games and Autism Special Education

We humans use gestures to express and communicate. Gesture games take the advantage of its easy and natural way for gamers to play. Combining visual, audio, and proprioceptive and motion controls, gesture games are increasingly used today in education or entertainment due to the fun element embedded. Research from Nottingham Trent University [7] found that Gesture Games can improve students with disabilities in their individual abilities on many aspects. In their research, after participating in game therapy, 92 % students with disabilities expressed interest to use such games in their future school learning. 92 % of them believed that they could learn more efficiently through game-based instructions rather than traditional learning ways.

Suzhou Industrial Park Renai School offers special provision for children with development disabilities. A number of children in the school are diagnosed as autism with various learning difficulties. For example, they may have sensory integrative dysfunction (SID), neuropsychological dysfunction, deficits in communication skills, deficits in social interaction skills, and a series of behavior issues. As a global trend, autism is having a significant increase recent years [8]. According to the latest figures released in March 2013, the incidence of autism amounts 2 % that means every 50 people there will likely be one person diagnosed with autism. It is internationally recognized that the best way to help children with autism is to provide them appropriate education. Until now, however, there are very few educational solutions effective for treatments of autism.

Renai School believes that to educate children with autism through play or game-based therapy would assist them to improve their social developments. The school also is willing to learn and share prospective practice good to children with autism. No doubt play or game-based therapy is a low cost yet high efficient way for autism education which is easy to popularize. First, there is a good potential to develop rich content for play or game activities which are not difficult to organize. It might be even possible for a widespread acceptance by each and every autistic child. Second, through game play, children with autism may become better motivate to interact or communicate with others. Third, for those with severe disabilities in speech and language, play would be an alternative way to foster their communication and interaction with other children. Teachers who play with those children may obtain more useful and 'inside' information from the play. Finally, we think play should be part of life for children with autism, and play therapy therefore should become a long-term strategy for autism education. In summary, we firmly believe that play or game-based therapy would bring children with autism more positive effects on verbal communication skills, better motion imitations from surroundings, and better peer interactions with classmates.

2 Theories of Gesture Games for Autistic Special Education

One of the major aims for autism education is to promote social interaction, verbal communication, and positive behavior. For children with autism, first they are children, and second they are children with disabilities. According to Montessori teaching theory [9], it is highly recommended to educate these children using play or game strategy which will help promote their social adaptabilities.

Gesture games not only have the similar functions of normal games, but also are fun filling due to the interaction using the natural user interface. Often, game play sessions can be conducted any places from classrooms to home. As games are designed for computers desktop, laptop, or portable devices including iPADs or smartphones, game-based learning is feasible anytime and anywhere. Specifically, Gesture Games can be used to support education at school or home as well as rehabilitation for children with autism. Gesture Games rely on high-tech video motion capture techniques. Body action can have immediately responses in the game system through the induction of human body movement. This real-time interaction promotes the game play especially for learning application. Cameras are used to capture the player body movements or gestures, after some data analysis, to start human–computer interactions [10]. Compared with traditional role play or video games, Gesture Games have advantage in the use of natural interaction for learning. Such interaction is difficult to have with the traditional mouse and keyboard. Gesture Games are often designed with advanced features for man–machine interaction, competition, and cooperation providing entertainment experience. These functions are suitable for children with autism in effective education, and educational rehabilitation.

2.1 Human–Computer Interaction

Interaction is a process of communication between two or more parties. Students interact with their peers or teachers in classrooms. They can also interact with objects such as books and footballs. In game design, players can interact with the avatars (peer students or teachers), books, and footballs in the virtual schools. In

Gesture Games, human–computer interaction is available where students (human) serving as players interact with the computer through avatars or virtual objects in the games using gesture control.

The recent basic curriculum reform in China emphasizes on the interactions between teachers and students, interactions between students and students, as well as interactions between human and computers under the ICT (Information and Communication Technology) environment. Human–computer interaction is one of the most attractive features in the game. Gesture Games open new avenues in learning by encouraging interaction through motion tracking and recognition in three-dimensional space and in real time. Gesture game technology is changing way of gaming by introducing new game experience. It enables children to participate in the game using their natural body movements. When Gesture Games are used for learning, students can have their scorings recorded to show their interactive learning progress. Human–computer interaction in the Gesture Games for children with autism has the potential to motivate and assist them in learning, and eventually, to enhance their social interaction.

2.2 Competitive Cooperation

In the process of cooperative learning, students may enhance their ability of communication and thus develop preliminary social skills. Gesture Games not only can engage children with autism in gameplay, but also can provide them a way to adapt their behavior in relation to team they play with. During the competitive game play, children with autism need to collaborate with their game partners. Team work thus becomes naturally embedded in the play sessions. For instance, they can do interact with each other face to face with corresponding body movement. From a long-term perspective, through game play in the Gesture Games children with autism may gradually change their behavior, become more positive and open minded. As such, cooperative game play using Gesture Games has a potential as part of educational rehabilitation to improve the learning of children with autism and reduce their rigid, compulsive, and stereotyped behaviors.

2.3 Emotional Learning

Emotional disability is one of major problems for children with autism. Gesture Games may be designed to have functions enabling emotion learning. For instance, Gesture Games can provide colorful pictures, vivid sounds, and realistic scenes. Advanced 3D technology is able to offer excellent stereo sensory experience making the game players feel real and act real. This may help to stimulate children with autism thus improve their emotion experience. By participating in Gesture Gameplay, children with autism may learn to improve their compliance, enthusiasm, and initiatives through a fun-filling experience. As such, Gesture Games can be used as emotion learning tool to assist these children better developing their social communication and adaptability.

3 Educational Practice of Kinect Gesture Game in Renai

Popular in western countries, Gesture games were introduced to China only recently. Kinect with motion tracking function was launched in November 2010 by Microsoft. It is able to capture, track, and recognize human body movements, including gestures and voices. Players can stand in front of the Kinect device to interact with the computer through a friendly and natural user interface. Kinect-based Gesture Games allow players directly use their body movement plus voice to control the games. A good feature with Kinect is that players can easily acquire the joyful experience through simple operation with the sensor. This kind of Gesture Games is being rapidly applied in education taking the advantages of their interactive, interesting, and entertaining features. Accordingly, we start our Kinect Gesture Games project in our school to explore its application on educational rehabilitation for children with autism. Our strategy focuses on the cognitive development [11] for autistic children with the optimally designed teaching plans through appropriate training as well as analytics based on feedback data. We will present the following four major parts which we are currently carrying out on our educational practice of Gesture Games.

First, we evaluate the current Kinect Gesture Games developed by Microsoft. Next, we will do a classification according to the Gesture Games we analyzed. We will then design our classroom practice with selected Gesture Games identified taking into consideration of our students' interests and abilities. Integration of various resources is finally done for curriculum based teaching and learning.

3.1 The Subject

Born in 2003, female student Julie has been diagnosed with autism. Her mother had normal pregnancy before giving birth to Julie. She did not show any delay in physical development ability except her social cognitive development including speech and language deficit. In February 2014, she carried a social life ability assessment combined with a development assessment. Evaluation results showed that Julie's independent life ability is about of 6-year-old children, athletic ability of 3.5-year-old, work skills of 6 year-old, social interaction skills of 3-year-old, ability to participate in group activity of 2-year-old, and self-management ability of 2-year-old. And through the developmental evaluation, the psychological consultant teacher found that generally Julie's cognitive development stage is around the age of 4. According to the above assessment results, Julie has a good performance

Table 1 Evaluation of selected Kinect Gesture games

Game title	Game evaluation	Category	Difficult level
Fruit Ninja	Simulations: free fruit cutting with hand gesture Functions: attention, coordination and upper limb movement	Casual puzzle	★
Big Adventure	Simulations: body center of gravity moving ships turn around, jump over the obstacles Functions: attention, balance coordination, partners	Action	★
Disneyland	Simulations: walk, turn, and say "hello" Functions: speech communication, social interaction	Role play	★★★
Motion Sports	Simulations: glide, biking, and rock climbing Functions: exercise fitness, jumping, onset of coordination	Sport	★★★★
Just Dance Now	Simulations: rhythm dance accompanied by music Functions: rhythm, physical coordination	Action	★★
Your Shape	Simulations: hands up the balance sheet, keep balance Functions: attention, balance and emotional control	Sport	★★
Puss in Boots	Simulations: move left to right, avoid obstacles ahead Functions: attention, reaction, sense of direction	Action	★★

in the independent living, but extremely weak in social and cognitive ability. In addition, she prefers to play alone, and is not sensitive to the environment change around her. At the same time, she has difficulties in communication or understanding, shows little interest, and often carries severely emotional problems. So she hardly understands the rules of the game too.

This case study uses classroom observation, parents and teachers interviews. It demonstrates four main aspects relating to our educational practice of Gesture Games for children with autism in our school (Table 1).

3.2 Lesson Plan and Participation

Children with autism often avoid eye contact with others. But this does not necessarily mean they do not have a common attention ability. Proper intervention with the aid of sensory-based interaction using Gesture Games has the potential to improve their joint attention through their eyes. The Kinect-based Gesture Games provide a convenient human–computer interaction with easy changes of scenes, and colors stimulating children with autism to keep common attention in a longer time during a gameplay session.

Focusing on the cognitive understanding, a lesson plan was designed for Julie in the case study.

- Gesture Games: "Fruit Ninja" and "Big Adventure".
- Duration: 30 min each lesson.
- Lesson Comparison: Art lesson, Math lesson, and Gesture Games (GG) lesson.

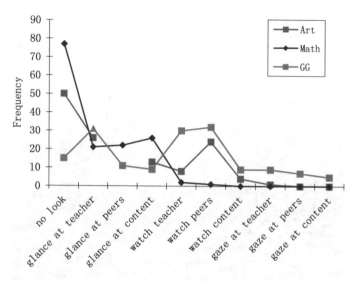

Fig. 1 Comparison of the gesture games versus other learnings

- Teacher's role: Facilitation and observation.
- Video Recording: Storm video software with the digital camera is used to record Julie's play at an interval of 10 s.
- Photographing: Remote mode with a distance 22 times the speed of the focal length.
- Analysis: Data recorded in the computer is analyzed, and observation analysis result is shown in Fig. 1.

From the analysis of the observation data in the graph by comparing Art, Math, and Gesture Games lessons, we found that Julie has improved significantly in the ability of 'glance at teacher,' 'watch teacher,' and 'watch peers'. The number of times of those abilities has increased obviously, and she has developed much more joint attention in her Gesture Game lesson.

Peripherals of Kinect Gesture Game are equipped with real-time dynamic capture, image recognition, speech recognition, social interaction, and other functions. Computer will identify, analyze, and according to a scheduled sensor model, to feedback in the computer side accordingly with the information come from human senses of sight, hearing, touch, and motion.

3.3 Coordination

Figure 2 shows Julie in the middle of playing the Gesture Game "Fruit Ninja". Importantly, to complete the game, she manages to coordinate her hands, eyes, and body movements. She was comfortable standing in front of the digital camera, and

Fig. 2 Julie performed in the Fruit Ninja gesture game

other recording device yet focus on her interaction with the Kinect-enabled "Fruit Ninja" game. Julie played the game role "Ninja" using a natural interface to trigger the progress of the game through an interactive feedback in terms of sound, light, electricity, and a variety of media of mutual stimulations.

The use of natural senses and auxiliary instructions tools help the coordination during the game play via smooth and continuous human–computer interactions. Throughout the rehabilitation process, it is also important to have some supporting elements implemented such as realistic context scenarios, lively cheers by fellow students, encouraging signs of success which can help children with autism gain "immersive" feeling. This, in turn, promotes human–human interaction.

3.4 Learning by Playing

Kinect sensor can actively track players' body movement, extraction players' body skeleton, and recognize the gestures within the sensor range. Kinect sensor is suitable to promote communication between students and teachers, as well as to enable students interacting with objects in a virtual space. These are very important in learning. Kinect Gesture Games we chose are able to provide visual realism, and auditory and other sensory feedback for teaching and learning.

The game 'Your Shape' is another example for children with autism to learn through play. Using Kinect motion control, players can put stones into a box, the more gem they collect, the higher they score. During the game, the computer system would show the pose and motion image of players. Children with autism would learn balance adjustment on their shoulders, elbows, and heads on their own based on the feedback they have received, so as to improve their balancing ability.

3.5 Social Communication

For Kinect Gesture Games, its peripherals may be more and more comprehensive that makes the human sensory multimodal interaction consistently improving. Teachers and students could not only get systemic participation on ideas or body movements, but also support many forms of interactive entertainment. During the process of gameplay, peers can work together and share their experience via verbal or nonverbal communication. Moreover, the Gesture Games as a media can help promote cooperation and team work in addition to competition. In the process of the play, students, teachers, and parents could share their emotions in a group to improve the ability of children with autism in speech, language, and social communication in a fun way.

In Julie's case, she participated in the Kinect Gesture Games with the supports from her parents. In May 2014, we conducted a preliminary interview with Julie's mother. She was briefed on the Gesture Game sessions Julie is going to attend. Her feedback was taken into consideration to develop a modified version of the lesson plan and other teaching practice for Julie. One month later, we carried out a formal interview with her mother lasting about 1 h. With consent from Julie's mother, here are parts of reflect from the interview after the game sessions.

1. Since taking part in the Kinect Gesture Games, Julie showed a significant change to her attention toward the surrounding world. At the same time, her understanding of the rules of the game has been improved slightly too. She is not much sensitive to the sound around her and develops the tolerance to the external environment.
2. The Kinect Gesture Game largely stimulated Julie's learning interest. Now she can focus better and keep longer time on the game she wants to learn how to play. It indeed inspired her learning enthusiasm.
3. When Julie began to participate in the virtual reality learning, she could understand the scene and the rules of the game as fast as it were. She can interact not only with the computer, but also with peers in the game. She even showed surprisingly her initiative to help her classmates on gaming. We also found that her language communication ability has improved. For instance, she occasionally tried to say 'I want to XXX' for a demand expression. She has decreased gradually her emotional problems such as the number of temper tantrums and manic shouting.

The analysis based on observation data of Julie's gameplay and feedback from her parents and teachers, we are delighted to see that the Kinect Gesture Game is playing a significant role in enhancing the social development of children with autism in the ability of verbal expression, communication skills, partnership, and interest behavior.

4   Conclusion

Nowadays, Information and Communication Technology (ICT) has penetrated into all aspects of the special education field. The environment of ICT-enabled educational rehabilitation for children with autism is rapidly improving. Chinese Education Information Development Plan For Ten Years (2011–2020) encourages "to promote the information technology and teaching integration, to help all school-age children and adolescents to use information technology with equality, effective and healthy; to provide information terminal equipment to meet the requirements of the learning disabled students and high-quality digital education resources."

People-oriented technology is experiencing a rapid development which has a good application in special education. Suzhou Industrial Park Renai School takes the lead in using the Gesture Games as innovations on the educational rehabilitation for children with autism. Some promising results have been obtained so far. In future, we will spare our efforts along this direction by developing more appropriate Gesture Game resources, establishing a new educational rehabilitation mode for autistic children based on the Gesture Game, sharing related findings in the research practice, and striving to make due contribution to the development of special education in China and in the world.

Acknowledgments This project is supported by Ministry of Education, China, under the "12–5" National Education Science Plan. The authors would like to thank Seagate China Charity Golf for their sponsorship with the project Star House. Thanks should also go to Nanyang Technological University, Singapore, and Singapore Millennium Foundation for their support in this research.

References

1. Pastor, I., Hayes, H.A., Amber, S.J.M.: A feasibility study of an upper limb rehabilitation system using Kinect and computer games. EMBC **28**(1), 1286–1289 (2012)
2. Lange, B., Chang, C.Y., Suma, E., et al.: Development and evaluation of low cost game-based balance rehabilitation tool using the Microsoft Kinect sensor. In: Proceedings of Conference on IEEE Engineering in Medicine and Biological Society, pp. 1831–1834 (2011)
3. Lin, Y., Shi, R., Chang, C.: Global community cloud learning. The 16th World Chinese Computer Education with Assembly (GCCCE2012) Review. J. Dist. Educ. **2012**(5), 13–19 (2012)

4. Cao, X., et al.: Gesture game application in stroke rehabilitation. J. Jilin Inst. Med. **2014**(3), 12–13 (2014)
5. Ma, J.: Parent–Child interaction based on the technology of gesture game design and implementation. China Educ. Technol. **9**, 85–88 (2012)
6. Liu, F., Shi, T.: Gesture game role of adolescent exercise behavior research. Bull. Sport Sci. Technol. **2013**(5), 26–28 (2013)
7. Nottingham Trent University.: Computer games could help people with learning difficulties to master everyday tasks. http://www.ntu.ac.uk/apps/news/109137-22/Computer_games_could_help_people_with_learning_difficulties_to_master_every.aspx (2011). Last visit on 27 May 2015
8. Chinese Radio Network.: Trends in Autism. http://native.cnr.cn/city/201304/t20130403_512285767.shtml (2013). Last visit on 27 May 2015
9. Maria, Montessori: Montessori Education Series in Chinese 蒙特梭利教育丛书. Translated by Wei Miao, Jilin Education Publisher (2012)
10. Dai, A., Qu, C., et al.: Gesture game interaction technology in the application of exercise rehabilitation leadership. Chin. J. Rehabil. Theory Pract. **2014**(1), 41–45 (2014)
11. Zhou, N.: Back of Autistic children's cognitive development research and exploration. Chin. J. Spec. Educ. (2002)

The Effectiveness of the Game LINGO Online: A Serious Game for English Pronunciation

Wim Trooster, Sui Lin Goei, Anouk Ticheloven, Esther Oprins,
Gillian van de Boer-Visschedijk, Gemma Corbalan
and Martin Van Schaik

Abstract In this chapter an evaluation of the LINGO Online, a serious game for English pronunciation is described. The game is applied in two primary and two secondary schools in the Netherlands. During 8 weeks of experiment, students from the schools either played the game or followed regular English lessons. Data were collected on learning outcomes (pronunciation performance), student learner characteristics, game characteristics, expectations and experiences of teachers and their coordinators, and teaching practice. Results on the effectiveness of the game are summarized (with clues for the working mechanism of the game). It shows that the game has the potential to compensate for insufficient facilities for English pronunciation education, to save teaching time, and to make the student less dependent on the expertise of the teacher. However, implementation of the game introduces new (technical) challenges.

Keywords English pronunciation · Serious gaming · Student learner characteristics · Game characteristics · Learning outcomes

W. Trooster (✉) · S.L. Goei
Windesheim University of Applied Sciences, Zwolle, The Netherlands
e-mail: W.Trooster@windesheim.nl

S.L. Goei
VU Amsterdam, Amsterdam, The Netherlands

A. Ticheloven
Windesheim University, Utrecht, The Netherlands

E. Oprins · G. van de Boer-Visschedijk · G. Corbalan · M. Van Schaik
TNO, Soesterberg, The Netherlands

© Springer Science+Business Media Singapore 2017
Y. Cai et al. (eds.), *Simulation and Serious Games for Education*,
Gaming Media and Social Effects, DOI 10.1007/978-981-10-0861-0_9

Abbreviations

I.T. Information technology
OMT One minute test
VAST Voice activated spy tech

1 Introduction

In Dutch education, more and more curriculum time is spent on English language. Dutch students used to have their first lessons on the English language in secondary education (starting at the age of 12). Nowadays, pupils already have English lessons in primary schools, at the age of 10 [29]. There are primary schools offering English lessons even for pupils at the age of 6 [6, 21]. Recently, the Dutch Ministry of Education [5] issued legislation that offers opportunity (not obligation) for primary education to use English language as the leading language during part of the time in the curriculum on English language.

At this moment, focus is primarily made on communication in English language lessons. English pronunciation is not an official curriculum topic in The Netherlands [10, 24]. Therefore, as only in a minor part of English lessons, and English pronunciation is conducted mostly in groups, in which the students are asked to practice via verbal phrases in an interview format as prescribed by the textbook used. Little attention is paid to pronunciation at the individual level. Once or twice a year students give an oral presentation in English. Only on these occasions teachers give feedback on the individual student's English pronunciation.

While learning correct English pronunciation is important; teaching time spent on English pronunciation remains limited due to the packed curriculum. Another bottleneck is that teachers, especially those at the primary school level, often do not have the skills to evaluate English pronunciation and to deliver appropriate feedback [19].

This study focuses on the role of e-learning in educational innovation. Information Technology could help the use of available curriculum time more effective and efficient [1, 9, 17, 25, 26]. E-learning using IT has the potential to overcome the problems described above.

In fact, software for personal computers and mobile devices is already being used in language courses [2, 3, 12, 18].

What relatively new is the use of games in education [31]. The so-called serious games are effective in a broad domain of skills, cognition and attitudes [4, 7, 15, 16, 30]. Serious games are digital games that are developed not only for fun and entertainment, but also for educational purposes, in which an interactive aspect can be implemented easily [26, 27, 31]. Previous studies have shown that serious games are effective in language courses [22] and they can be used to evaluate the performance of individual students and generate feedback for each of them [27], in the fields of writing [23], English grammar [13], and usage and vocabulary [13, 32]. As such serious games might be an effective tool in teaching English pronunciation in second language education.

The scanty attention paid to the teaching of speaking skills could also be due to the complexity of assessing and evaluating them, which is more ambiguous and time consuming than the assessment of other skills. And teaching speaking, especially pronunciation skills, to a large classroom is normally more difficult and time-consuming than teaching other skills. Serious games can have the potential to overcome problems like these, because games can be played individually and games offer instant and specific feedback [27].

In Singapore Magma Studios has developed the LINGO Online (formerly called the VAST-game: "*Voice Activated Spy Tech*"), an online multiplayer serious game for desktop/laptop. It offers 40 h of training in English pronunciation [14]. The game was designed for students in the final grades of primary education and students in the first grades of secondary education (ages 10–16) in Singapore. In this game students are challenged to pronounce texts presented by the game. The game characters speak in text and voice simultaneously. The game software evaluates the students' pronunciation and offers instant feedback in response (see Fig. 1).

The aim of the study presented in this chapter is to determine the effectiveness of the game LINGO Online for development of accurate English pronunciation by Dutch native speaking students in the Dutch educational context. To study this, students in primary and secondary education were trained in English pronunciation

Fig. 1 Screenshot from the game LINGO Online. The game offers instant feedback on the student's pronunciation: text in *green* was pronounced correctly, text in *orange* was dubious, text in *purple* was incorrect. The student can listen to a record of his/her pronunciation (*green button*) and to pronunciation by a native speaker (*button with teacher image*)

by means of LINGO Online. Their performance was compared with the performance of students in primary and secondary schools who had lessons in English as usual. In addition, the educational context was analyzed to determine relevant contextual parameters that could influence the game's effectiveness.

1.1 Research Questions

The main research question of this study reported in this chapter is as follows:

How effective is the serious game LINGO Online and how is the game used in the Dutch educational context?

In order to study this main research question, the following three sub-questions are addressed:

RQ 1. *Does playing LINGO Online improve the learning of English pronunciation, as compared to not playing with the game?*

RQ 2. *Does playing LINGO Online yield higher scores on factors that positively influence the learning process, as compared to not playing with the game?*

RQ 3. *Which game characteristics possibly contribute to results for RQs 1 and 2?*

In addition, data regarding the conditions under which participating teachers implemented the game were also gathered in order to better understand the context under which learning with the game took place. This information is not gathered systematically and is used as illustration and to interpret the results of *RQ*s 1, 2 and 3.

2 Methods and Materials

During the period September 2013–January 2014, students and teachers from two primary schools (each two groups; students in the final two grades, age range 9–13 years old), and students and teachers from two secondary schools (each two groups; students in second and third grades, age range 13–15 years old) participated in the study.

To evaluate the effectiveness and mechanism of action of LINGO Online, a methodology and research tools were elaborated based on the methods developed in prior research [20]. During 8 weeks in each school, 1 group of students ($n = 25$, approximately) played the game (Game condition). A parallel group followed regular lessons (Control condition "Regular lessons" without the game). Data for the study were collected before (pretest), during and after (posttest) this 8-week period. Figure 2 shows the types of data collected in the pretest, posttest and during the 8-week period.

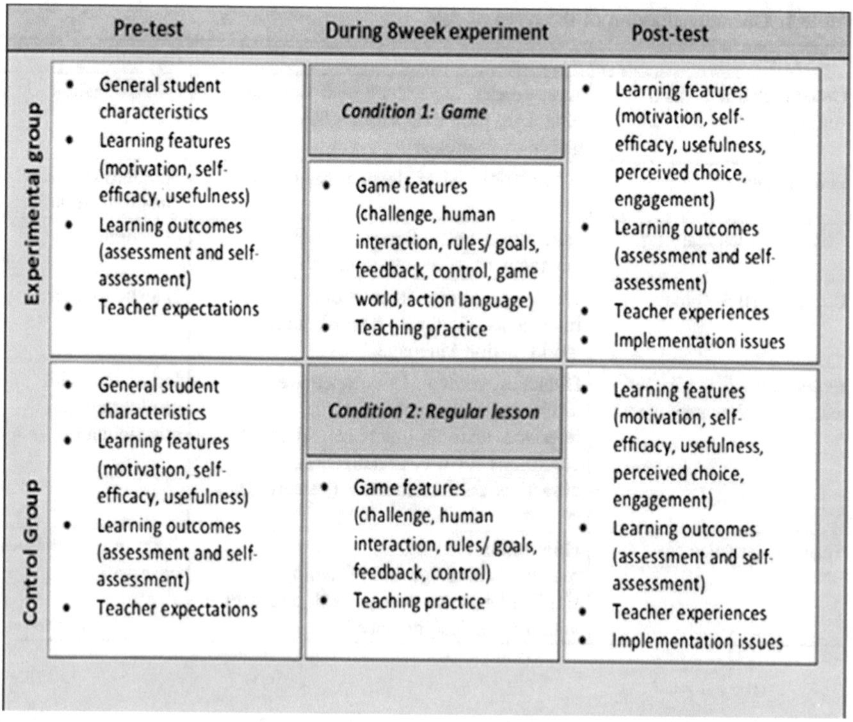

	Pre-test	During 8week experiment	Post-test
Experimental group	• General student characteristics • Learning features (motivation, self-efficacy, usefulness) • Learning outcomes (assessment and self-assessment) • Teacher expectations	**Condition 1: Game** • Game features (challenge, human interaction, rules/goals, feedback, control, game world, action language) • Teaching practice	• Learning features (motivation, self-efficacy, usefulness, perceived choice, engagement) • Learning outcomes (assessment and self-assessment) • Teacher experiences • Implementation issues
Control Group	• General student characteristics • Learning features (motivation, self-efficacy, usefulness) • Learning outcomes (assessment and self-assessment) • Teacher expectations	**Condition 2: Regular lesson** • Game features (challenge, human interaction, rules/goals, feedback, control) • Teaching practice	• Learning features (motivation, self-efficacy, usefulness, perceived choice, engagement) • Learning outcomes (assessment and self-assessment) • Teacher experiences • Implementation issues

Fig. 2 Types of data collected in the pretest, posttest and during the 8 week period

Performance was assessed with the standardized English One Minute Test (OMT) [8]. It measures general fluency of word reading in English. Here the score for each student was the number of words (from a list of frequently used English words) read and pronounced correctly during 1 min [11]. Furthermore, each student was asked to read aloud the Tommy text, a simple text of 228 words, containing several words that occurred in the game that was tested in this study. In contrast to the OMT, where words are presented in isolation (context-independent), in the Tommy Text each word is embedded in a meaningful context. The participant's score was the number of words pronounced in correct Oxford English. The total number of words read correctly per minute was also calculated. In Table 1 the other data types are operationalized; for more detail see the technical report on the study by Trooster et al. [28].

From the raw data on those parameters examined in both the pretest and the posttest delta's (difference posttest–pretest) were calculated. These delta's and the data on the remaining parameters were compared between the experimental and the control conditions by using *t*-tests.

To explore the mechanism of action of the game LINGO Online, correlations were analyzed between the four types of parameters (learning outcomes, learner

Table 1 Operationalization of the types of data

Data type	Examples	Instruments
General characteristics students	Age, gender, experience with English language, problems learning/reading, gaming experience	Questionnaires
Performance	Articulation, word stress, tone, fluency	Standardized tests questionnaires
Characteristics learning	Motivation, self-efficacy, control, engagement, concentration, value	Questionnaires
Characteristics game	Challenge, human interaction, rules/goals, feedback, control, game world, action language	Questionnaires
Expectations/experiences teachers and/or managers	Didactics, content, IT competence teachers, facilitation, technique, interface, learning outcomes, adaptivity, conditions for use, organization, attractiveness for students, (perceived) success	Interviews with semi-structured questionnaires
Teaching practice	How often reading/writing/speaking/listening English language during week and with which didactics/guidance	Logs teachers observations

characteristics, game characteristics and parameters on the context of learning). In case of a significant correlation this analysis was followed by regression analysis.

3   Results

Data have been collected from two primary schools and two secondary schools (each school ±25 students in gaming group and ±25 students in control group). One secondary school did not succeed in implementing the game (technical installation on the computer network failed, in spite of external support). In this school data were collected only from the teacher and coordinator (not from the students).

In the remaining three schools all data on the relevant parameters were collected, processed (calculating delta's) and analyzed (comparisons, correlations and regressions): data on speech performance of the students (combined with their general characteristics), learner characteristics of the students, game characteristics, expectations and experiences of teachers and their managers, and data on their teaching practice. Data from the language tests are key to describe performance on English pronunciation (and thus effectiveness of the game: learning outcomes). Especially collection and analysis of these data on pronunciation performance proved to be very time-consuming.

As a first impression of the potential of this serious game, teachers from the schools where LINGO Online was running indicated that students playing the game felt more free to speak English than students following the regular lessons. Furthermore, these teachers indicated that the students playing the game were highly motivated: even though these students frequently encountered (multiple) technical problems (they sometimes had to wait for up to 15 min) these students persisted in playing the game.

In the next sections the results are described: on the learning outcomes, learner characteristics, game characteristics and the correlations between these parameters.

3.1 Learning Outcomes

In both the game condition and the control condition the OMT-scores were significantly higher in the posttest than in the pretest. For the primary school students the delta scores (difference posttest–pretest) of the OMT were significantly higher in the game condition than in the control condition. By contrast, for the secondary school students the delta scores of the OMT in the game condition were significantly lower than in the control condition.

For the Tommy test the number of pronunciation errors and the time needed to read the text were scored. For both the game condition and the control condition the number of errors and the reading time decreased significantly from pretest to posttest. However, the delta scores did not differ between the two conditions (gaming vs. control).

During observations teachers mentioned that the foremost advantage of the game was to lower the threshold to speak English aloud in the classroom. The data from the self-assessments from primary school students show the tendency that fluency of speaking increased from pretest to posttest in the game condition. However, a similar increase was found in the control condition.

No significant differences between pretest and posttest were found for the competences to use the correct articulation, tone and word stress.

3.2 Learner Characteristics

All students completed questionnaires on their learner characteristics.

For the game condition motivation increased significantly from pretest to posttest. For the control condition there was no significant difference between pretest and posttest. Perceived Choice was scored in the posttest, not in the pretest. The mean score for Perceived Choice was significantly higher in the game condition than in the control condition. No significant differences were found between the game condition and the control condition for the learning characteristics self-efficacy, value, engagement.

Motivation and engagement correlated with the learning outcomes scored with the OMT and Tommy Test (number of errors). No correlation was found between self-efficacy, perceived choice or value and these learning outcomes.

3.3 Gaming Characteristics

All students completed questionnaires on the gaming characteristics they experienced during the lessons. The mean scores of the gaming characteristics feedback, challenge & control and rules & goals did not differ for the game condition and the control condition. The game characteristics Game World and Action Language are specific for the game condition (not relevant in regular lessons) and were not included in the comparison here.

No significant correlations were found between Gaming Characteristics and Learning Outcomes. All five gaming characteristics correlated with the learning feature motivation. The gaming characteristics Feedback and Rules & Goals correlated with all learner characteristics.

3.4 Issues During Implementation of the Game

All schools experienced critical issues during implementation of the game. These issues challenged the motivation of teachers and students. Before using the game, installation of the Flash-player and Firefox-browser was necessary. Teachers had questions where to download this software and network security prevented installation of the software downloaded by the teachers. Firewall settings needed to be changed to enable traffic of speech produced during gaming over internet to the server. Teachers were not able to make these changes. In the configuration settings of the individual computers software drivers for headsets had to be selected and software drivers for the speakers of the computers had to be deselected. Again, teachers were not able to make these changes. Although the teachers instructed the students how to use the game (with a manual), students still had questions on how to use the game.

Giving extra hints and tips the teachers and fellow students helped students with questions on gameplay. Questions on technical issues needed support from technical experts.

In one secondary school sufficient technical support was available. In this school implementation was successful. In the other secondary school technical support provided by network manager was inadequate, and support by the research team was not possible. As a consequence adequate installation of the game was not possible. In this school the pilot was cancelled. In the two primary schools no technical support was available. Here the research team provided the necessary support. This implementation of the game was successful.

4 Conclusion and Discussion

In the present study we investigated the effectiveness of the serious game LINGO Online by comparing students playing the game with students having regular English lessons over a period of 8 weeks. In order to study this main research question, three sub-questions were addressed:

1. Does playing LINGO Online improve the learning of English pronunciation, as compared to not playing with the game?
2. Does playing LINGO Online yield higher scores on factors that positively influence the learning process, as compared to not playing with the game?
3. Which game characteristics possibly contribute to results for RQ 1 and 2?

Also information was gathered regarding the conditions under which participating teachers implemented the game as to better understand the context under which learning with the game took place.

Does playing LINGO Online improve the learning of English pronunciation, as compared to not playing with the game?

For the two primary schools, the game had a significantly larger learning effect than the regular lessons in terms of automatized direct word identification of context-independent English words. For the secondary school, both groups (control and experimental) also learned, but when looking at the gains, there is less growth for students playing the game compared to those having the regular lessons. Also, for pronunciation errors and reading time, as measured by reading aloud a short text; again the primary schools showed greater progress than the secondary school. It is concluded that playing LINGO Online improves the learning of English pronunciation as compared to not playing the game, especially in primary education.

Does playing LINGO Online yield higher scores on factors that positively influence the learning process, as compared to not playing with the game?

The most noteworthy result is for the learning characteristic 'motivation'. The difference between the two groups for this parameter is significant, indicating that the game group is more motivated than the control group. Perceived choice scored significantly higher in the game group compared to those having the regular lessons. In other words, the students who played the game had the feeling they had more influence on what they did during the lessons and felt less obliged to participate compared to the students from the control group. It can be concluded that playing LINGO Online yields higher scores on the factors Motivation and Perceived choice (but not Value, Engagement and Self-efficacy), as compared to not playing the game. Looking at the correlations between the learner characteristics and the learning outcomes, it is suggested that motivation and engagement, in particular, influence the learning outcomes.

Which game characteristics possibly contribute to results for RQs 1 and 2?

For the game characteristics (Feedback, Challenge & Control, Rules & Goals, Action Language, Game world) no significant differences were found between the game group compared to those having the regular lessons. So we cannot say that the game is definitely better designed than regular lessons from an educational design point-of-view. No significant correlations were found between game characteristics and outcome measures. This shows that the game characteristics do not contribute to the learning outcomes directly.

It is very interesting to see that all game characteristics correlated with the learner characteristic motivation, and that the game characteristics feedback and rules & goals correlated with all learner characteristics.

This suggests that the game characteristics contribute to the learner characteristics. Through these learner characteristics the game characteristics may contribute to the learning outcomes indirectly.

In conclusion, the game LINGO Online proved to be of value as a learning tool for practicing English pronunciation, especially in primary schools. The data suggest a central role for motivation in how LINGO Online operates to achieve the learning objectives.

The game LINGO Online has the potential to compensate for insufficient facilities (time, expertise) for English pronunciation education in schools. Using the game the students are able to train English pronunciation individually. The students are motivated to learning in a self-directed strategy (thus saving time); using the learning analytics in the software or triggered by questions of students the teacher is able to deliver specific help to the individual students (making the student less dependent of the expertise of the teacher).

However, the game introduced new challenges menacing motivation of teachers and students. The schools experienced critical issues during the implementation of the game: installation of secondary software was necessary; firewall settings needed to be changed; software drivers for headsets should be selected; after instruction and with manual, students still had questions on how to use the game. These issues can be solved with adequate support.

Acknowledgments LINGO Online was created and developed by Magma Studios and Carnegie Speech Company utilizing game-based learning and speech analysis technologies. This product was originally created for the Ministry of Education in Singapore, known as V.A.S.T.: *Voice Activated Spy Tech*. It is now commercially available under the name of LINGO Online. For further information please contact: info@magma-studios.com, info@carnegiespeech.com.

Special thanks go to Anouk Ticheloven, who worked as a bachelor intern within the project, and onwards as a honour student in finalizing her bachelor thesis within the project.

This study was made possible by a grant from Kennisnet (www.kennisnet.nl) and internal funding from Windesheim University of Applied Sciences.

References

1. Blok, H., Oostdam, R., Otter, M., Overmaat, M.: Computer-assisted instruction in support of beginning reading instruction: a review. Rev. Educ. Res. **72**, 101–130 (2002)
2. Burston, J.: Mobile-assisted language learning: a selected annotated bibliography of implementation studies 1994–2012. Lang. Learn. Technol. **17**(3), 157–225 (2013)
3. Chinnery, G.M.: Emerging technologies: going to the MALL: mobile assisted language learning. Lang. Learn. Technol. **10**(1), 9–16 (2006)
4. Divjak, B., Tomić, D.: The impact of game-based learning on the achievement of learning goals and motivation for learning mathematics—literature review. J. Int. Org. Stud. **35**(1), 15–30 (2011)
5. Dutch Ministery of Education.: Aanbieden van onderwijstijd in de Engelse, Duitse of Franse taal voor het primair onderwijs. Website Dutch Senate, The Hague, The Netherlands. https://www.eerstekamer.nl/wetsvoorstel/34031_aanbieden_van_onderwijstijd (2015). Accessed 08 Oct 2015
6. Earlybird.: Landelijke ontwikkelingen. Website Earlybird, Rotterdam, The Netherlands. http://www.earlybirdie.nl/index.php?page=VVTO-Landelijke_ontwikkelingen&pid=184 (2013). Accessed 03 Apr 2014
7. Egenfeldt-Nielsen, S.: Overview of research on the educational use of video games. Digital Kompetanse **1**(3), 184–213 (2006)
8. Fawcett, A.J., Nicolson, R.I.: The Dyslexia Screening Test. The Psychological Corporation, London (1996)
9. Girard, C., Ecalle, E., Magnan, A.: Serious games as new educational tools: how effective are they? A meta-analysis of recent studies. J. Comput. Assist. Learn. **29**, 207–219 (2013). doi:10.1111/j.1365-2729.2012.00489.x
10. Greven, J., Letschert, J.: Kerndoelen Primair Onderwijs. Resource document SLO, Enschede, The Netherlands. http://www.slo.nl/primair/kerndoelen/Kerndoelenboekje.pdf/download (2006). Accessed 03 Apr 2014
11. Kleijnen, R., Steenbeek-Planting, E., Verhoeven, L.: Toetsen en Interventies bij Dyslexie in het Voortgezet Onderwijs: Nederlands en de moderne vreemde talen. Expertisecentrum Nederlands, Nijmegen (2008)
12. Kuang-wu, L.J.: English teachers' barriers to the use of computer-assisted language learning. Internet Teachers Engl. Second Lang. J. **6**(12), 1–8 (2000)
13. Kuppens, A.H.: Incidental language acquisition from television, video games, and music: an empirical study with Flemish youngsters. Paper presented at the annual meeting of the International Communication Association, Montreal, Quebec, Canada (2008)
14. Magma Studios.: LINGO Online: Speaking For Fun! Website MagmaStudios, Singapore. http://www.lingo-online.com/games/lingo-online (2014). Accessed 03 Apr 2014
15. Mayer, R.E., Johnson, C.I.: Adding instructional features that promote learning in a game-like environment. J. Educ. Comput. Res. **42**(3), 241–265 (2010)
16. Ming-Chaun, L., Chin-Chung, T.: Game-based learning in science education: a review of relevant research. J. Sci. Educ. Technol. **22**, 877–898 (2013). doi:10.1007/s10956-013-9436-x
17. Mioduser, D., Tur-Kaspa, H., Leitner, I.: The learning value of computer-based instruction of early reading skills. J. Comput. Assist. Learn. **16**, 54–63 (2000)
18. Mitchell, R., Myles, F., Marsden, E.: Second Language Learning Theories. Routledge, Abingdon (2013)
19. Onderwijsinspectie.: De Kwaliteit van Leraren. Resource document. Ministerie van Onderwijs, Cultuur en Wetenschap, The Hague, The Netherlands. http://www.onderwijsinspectie.nl/binaries/content/assets/Onderwijsverslagen/2012/ov1011_h9_kwaliteitleraren_printversie.pdf (2011). Accessed 03 Apr 2014
20. Oprins, E., Bakhuys-Roozeboom, M., Visschedijk, G.: Effectiviteit van serious gaming in het onderwijs. Onderwijsinnovatie, 32–34 (2013)

21. PO-Raad.: PO-Raad vooral positief over meer ruimte Engelse les in primair onderwijs. Website PO-Raad, Utrecht, The Netherlands. http://www.poraad.nl/content/po-raad-vooral-positief-over-meer-ruimte-engelse-les-primair-onderwijs (2013). Accessed 03 Apr 2014

22. Peterson, M.: Computerized games and simulations in computer-assisted language learning: a meta-analysis of research. Simul. Gaming **41**, 72–93 (2010). doi:10.1177/1046878109355684

23. Robertson, J., Good, J.: Using a collaborative virtual role-play environment to foster characterisation in stories. J. Interact. Learn. Res. **14**, 5–29 (2003)

24. SLO.: Europees Referentiekader. Resource document. SLO, Enschede, The Netherlands. http://view.officeapps.live.com/op/view.aspx?src=http%3A%2F%2F www. europeestaalportfolio.nl%2Fdocs%2FEuropees%2520Referentiekader.doc (2004). Accessed 17 Apr 2014

25. Schittek, M., Mattheos, N., Lyon, H.C., Attström, R.: Computer assisted learning. A review. Eur. J. Dent. Educ. **5**, 93–100 (2001)

26. Sørensen, B.H. and Meyer, B.: Serious Games in language learning and teaching–a theoretical perspective. In Proceedings of the 3rd International Conference of the Digital Games Research Association, Digital. Game. Res. Assoc. Tokyo Japan, pp. 559–566 (2007)

27. Sykes, J., Reinhardt, J.: Language at Play: Digital Games in Second and Foreign Language Teaching and Learning. Pearson, New York (2012)

28. Trooster, W.J., Goei S.L., Ticheloven, A.H.L., Oprins, E., Visschedijk, G., Corbalan, G., van Schaik, M.: The effectiveness of LINGO Online, A Serious Game for English Pronunciation. Unpublished Internal report, Windesheim University of Applied Sciences, Zwolle (2015)

29. Vedocep.: Status Eibo. Website Vedocep, Nijmegen, The Netherlands. http://www.vedocep. nl/pages/eibo.htm (2008). Accessed 03 Apr 2014

30. Vogel, J.J., Vogel, D.S., Cannon-Bowers, J., Bowers, G.A., Muse, K., Wright, M.: Computer gaming and interactive simulations for learning: a meta-analysis. J. Educ. Comput. Res. **34**(3), 229–243 (2006)

31. Young, M.F., Slota, S., Cutter, A.B., Jalette, G., Mullin, G., Lai, B., et al.: Our princess is in another castle: a review of trends in serious gaming for education. Rev. Educ. Res. **82**(1), 61–89 (2012). doi:10.3102/0034654312436980

32. Zheng, D.: Affordances of three-dimensional virtual environments for English language learning: an ecological psychological analysis. Dissertation Abstracts International Section A: Humanities and Social Sciences, vol. 67(6A), p. 2057 (2006)

A Gaze Tracking System for Children with Autism Spectrum Disorders

Yeli Feng and Yiyu Cai

Abstract Individuals with autism exhibit deficits in social communication and interaction. Researchers have been using interactive virtual reality (VR) technologies helping children with autism to improve their communication and learning. For example, children with autism interact with virtual dolphins using hand gesture, and learn social skills and safety skills. In this chapter, we propose a low cost training system aiming to enhance the visual responsiveness of children with autism. In the proposed system, children with autism use gaze to interact with tasks that are designed with game mechanics. The system can easily record data for analysis.

Keywords Eye tracking · Autism · Children · Game

1 Introduction

Since the early adoption of virtual reality (VR) technology in the research of autism spectrum disorder in 1996 [1], many research activities have been channeled towards using VR technology to understand the unique behaviors of autism; and to develop effective intervention program to treat autism. In 2002, Parsons and Mitchell [2] reviewed the potential usefulness of VR includes providing a safe and nonthreatening environment, a realistic settings for role-play of desired behaviors, can be repeatedly practiced in a consistent way. A decade later Parsons and Cobb [3], Bellani et al. [4] reviewed the state of the art in this field. They both concluded that children with autism can learn rule-based social skills from VR in varying degree and a few can transfer the acquired skills to real word. Other latest researches [5–7] suggest VR is promising in improving understanding of empathy and social interaction in children with autism. Our work focuses on the nonverbal

Y. Feng · Y. Cai (✉)
School of Mechanical and Aerospace Engineering, Nanyang Technological University, Singapore, Singapore
e-mail: myycai@ntu.edu.sg

© Springer Science+Business Media Singapore 2017
Y. Cai et al. (eds.), *Simulation and Serious Games for Education*,
Gaming Media and Social Effects, DOI 10.1007/978-981-10-0861-0_10

social interaction deficits in children with autism that using eye contact as a mean of communication.

This chapter introduces a low cost training system aiming to enhance the visual responsiveness of children with autism. Section 2 reviews the prior art. Section 3 presents the training system designed and developed for children with autism. Section 4 discusses data analysis methodology to evaluate the effectiveness of program. Section 5 concludes this research.

2 Literature Review

During the 1940s Leo Kanner and Hans Asperger independently reported children who presented "a powerful desire for loneness" and "an obsessive insistence on persistent sameness". Since then psychiatrists and psychologists looked into the life experience impacts for explanation and intervention for this behavior disorders in the domains of social deficits, communication, and language deficits. From the 1970s to 1990s many behavioral treatments based on the operant conditioning principles were developed targeting behaviors include social skills, language, daily living skills, academic skills, and aberrant behaviors [8]. Behavioral treatments systematically apply reinforcement, punishment, and extinction techniques or combined to develop effective treatment strategies. The trend continued to increase until today referred generally as applied behavior analysis (ABA) [9]. ABA and many variations are widely practiced in the field of autism and developmental disabilities in special education in US, UK, and many other countries.

In the 1985 article [11] "Does the autistic child have a 'theory of mind'?" Leslie suggested that the lack of 'theory of mind' is the underlying cognitive condition attributes to the impaired social communication in autistic children. The ToM is the ability to attribute beliefs, intentions, and other mental states to self and others in order to understand others' belief and intention. Later studies [2] indicate teaching ToM to people with autism is helpful to pass ToM tasks being received teaching but not new tasks. With the advance in nonintrusive corneal reflection eye tracking and the functional magnetic resonance imaging (fMRI) technologies, since early 2000s studies [12] have been carried out to search the characterization of eye movement of individuals with autism during social processing and the connection between the neurodevelopmental condition and behavioral dysfunctions. Evidences of atypical gazing pattern and hypoactivation in different nodes of the brain have been found although details varying in findings.

Over the time autism is being diagnosed more frequently. In 2010, the CDC monitoring network reports "The global prevalence of autism has increased 20-fold to 30-fold since the earliest epidemiologic studies were conducted in the late 1960s and early 1970s." Although the underlying reasons to the apparent high increasing rate are unclear, several factors likely contributed, such as improved awareness and services to general public, and changes in diagnostic criteria. The Diagnostic and

Statistical Manual of Mental Disorder, Fifth Edition (DSM-V) groups all subconditions such as Asperger syndrome into two categories: impaired social communication and or interaction, and restricted and or repetitive behaviors.

Gaze plays an important role in social interaction. Typically developed infants connect with caregivers in moments of gazing at caregivers' face, following caregivers' gaze direction. Around 1 year old they are capable of establishing joint attention. In everyday social processing typically developed adults frequently utilize eye contact signaling messages, such as friendship, agreement, or anger. One of the core symptoms of autism is the abnormality in eye contact with people and objects during social interaction.

In the autism research literature many studies investigated such abnormality. Using human faces as stimuli Pelphrey et al. [13] reported the scan paths of adult males with autism seemed erratic and disorganized. They had less fixations on the core features of the faces, and less accurate in identifying the six basic emotions fear in particular. Klin et al. [14] used naturalistic social situations to study the difference of fixation on social and nonsocial regions of a dynamic scene. They found adolescents and young adults with autism fixated significantly less on eye but more on mouth, body, and object regions when watch the videotape clips. Within the group autism participants with higher social competence had more fixations on mouth region. Von Hofsten et al. [15] reported children with autism had fewer fixations and averagely shorter fixation duration on the faces of the people involved in a conversation but no evidence of impairment on tracking moving objects. Chawarska and Shic [16] reported toddlers with autism distributed attention between inner and outer facial features differently than typically developed peers, hence are less effective in facial recognition. Attention to eyes, nose and mouth features decreased in elder toddlers with autism which suggested possible sign of detrimental over age.

3 Gaze Training System for Children with Autism

Autism is a neurodevelopmental disorder. People with autism have different hence less effective mechanism to gather information and communicate in the world favor of typically developed individuals. They experience different degree of difficulties in maintaining friendship, relationship, and employment which can lead to social exclusion. However, the development in the neuroplasticity field [17] gives hope to ABA intervention. The connectivity in the brain changes throughout the lifetime and is sensitive to the patterns produced through experience. A change in behavior signifies some change in the brain. Through ABA experience designed on repetition and reinforcement the emerging brain of children with autism is rewired and retrained to engage in behaviors that are socially better accepted.

ABA intervention requires expertise and long-term commitment. The accessibility and affordability could be the factors preventing it reaching children with autism in need. In this section, we propose a system that allows children with autism receive regular intervention without the need of ABA expertise always. The exercise can be carried out at school, home, or community facilities such as library. The system aims to help them enhancing visual responsiveness in the new media environment. First the enabling technology eye tracking is introduced. Then we describe the system framework and application development.

3.1 Eye Tracking Technology

Since eye tracking technology was first used in autism research [18] more than 10 years ago, tremendous improvements have been introduced [19]. Today most eye tracking devices used in the field of autism study fall into four categories. See Fig. 1 table-mounted with chin rest, head-mounted, glass-wear, and remote systems. These systems are video based, rely on the corneal-pupil reflection of infrared or ambient lights directed to the eye to estimate the point of gaze.

The corneal reflections are known as Purkinje images. Four Purkinje images are commonly used. The first and second Purkinje images are the reflections from the outer and inner surface of the cornea, respectively. The third and fourth Purkinje images are the reflections from the outer and inner surfaces of the lens respectively see Fig. 2. Gaze location is computed by extract the pupil center and one or more of the corneal reflections with image-processing algorithm. A calibration is required to map the estimated gaze location onto the viewing scene coordinates, for example x and y coordinates of the monitor [21].

We use Tobii EyeX Controller, a remote system that can be mounted on both desktop and laptop setups. It does not need regular recalibration and user can sit,

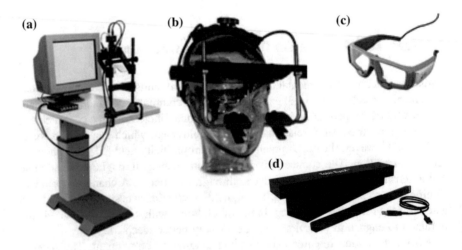

Fig. 1 a Cambridge research systems high-speed VET. **b** SR research EYELINK II. **c** SMI eye tracking glass. **d** Tobii EyeX controller

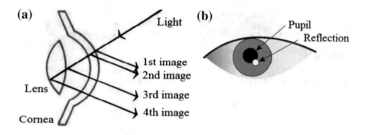

Fig. 2 **a** Purkinje images. **b** Illustration of pupil and reflection extracted by image-processing algorithm

stand, and move freely. The operating distance is between 45 and 80 cm, freedom of head movement is 48 cm width by 39 cm height at 70 cm distance.

3.2 System Framework

The gaze training system (Fig. 3) is built on top of the EyeX engine [22] and Unity [23] game engine. The EyeX engine filters the raw gaze data from the EyeX controller and reports up to gaze-enabled application or interprets into meaningful interactions which are defined at the application level as integrators. The Unity game engine is a cross platform 3D rendering engine comes with intuitive tools such as scene editor, scripting, and animation workflow.

As shown in Fig. 4 the gaze training system creates training tasks with the Unity game engine. The training tasks are simple games played via eye interaction. In

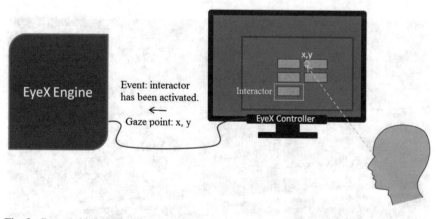

Fig. 3 Gaze-enabled application

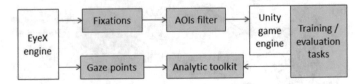

Fig. 4 Architecture of the training system

each training tasks areas of interest (AOI) are defined as interactors. Fixation that is gaze maintained on a single location for a longer time on the AOI triggers events that reward player. In the meantime the player's gaze data during entire session are captured for analysis.

3.3 Application Development

The training tasks we design and develop restrict children advance their game play with eye interaction only. In a training, task regions are defined to give reward, cue or punishment. For example in Fig. 5, during the play if the child fixes eye on the eyes of the dragons, he or she will be rewarded a coin. If the child fixes eye on the distractor a flyer, a visual cue that one of the dragons plays a cheer animation is given.

Studies [24] reveal that the gazing difference between typically developed individuals and with autism does not generalize across static and dynamic social and nonsocial stimuli. Therefore, two types of evaluation tasks are developed adopting widely used methods in the autism eye tracking research field. In paired comparison form sequence of static neutral face and scene of person-absent are

Fig. 5 Example of training task. During play mode the mask of *rectangle* regions and gaze and fixation tracker (*white* and *grey dots*) are turned off by default

Fig. 6 Example of evaluation task using static stimuli. *Line* depicts the scan path of gazing and *black squares* are the start and end points

flashed to viewer. Scan path, start, and end gaze points are captured for analysis, as shown in Fig. 6. Video clip of cartoon characters having conversation are used as dynamic stimuli, attention, and eye movement activities are captured for evaluation.

4 Data Analysis

Three types of playing data stream are captured; they are gaze points, fixation points, and triggered events. The triggered events data forms a profile of each child, able to win the game and how long it took. From the gaze points, we can infer whether the child with autism pays attention to the screen, how frequently he or she moves eye among different regions in the screen see Figs. 7 and 8. The fixation

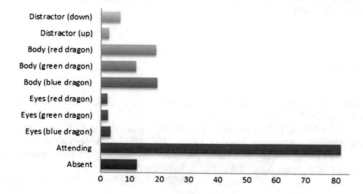

Fig. 7 Distribution of attention during training in seconds

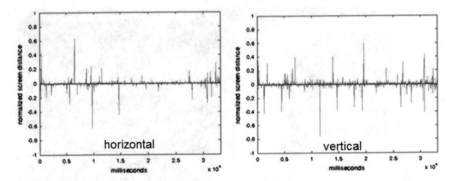

Fig. 8 Saccades of eye in an evaluation task

points reveal what features attract the player's attention if they are not the defined reinforcement regions. This type of information can be feedback to the design of training tasks, for example, adjust difficulty levels.

5 Conclusion

In this chapter, we discuss our initial work on gaze tracking for the possible use of game-based learning for children with autism. A data collection mechanism is designed and implemented with the system for analysis. The research can be applied to investigate the effectiveness and efficiency of using VR technology for learning [3].

Future work of this research includes the improvement of system design and evaluation. Object recognition, motion detection, and more general Artificial Intelligence will be studied for the purpose to develop better gaze tracking solutions.

Acknowledgments This project is partially supported by The Temasek Trust Funded Singapore Millennium Foundation.

References

1. Strickland, D.: A virtual reality application with autistic children. Presence-Teleoperators Virtual Environ. **5**(3), 319–329 (1996)
2. Parsons, S., Mitchell, P.: The potential of virtual reality in social skills training for people with autistic spectrum disorders. J. Intellect. Disabil. Res. **46**(5), 430–443 (2002)
3. Parsons, S., Cobb, S.: State-of-the-art of virtual reality technologies for children on the autism spectrum. Eur. J. Spec. Needs Educ. **26**(3), 355–366 (2011)
4. Bellani, M., Fornasari, L., Chittaro, L., Brambilla, P.: Virtual reality in autism: state of the art. Epidemiol. Psychiatr. Sci. **20**(03), 235–238 (2011)

5. Cai, Y., Chia, N.K., Thalmann, D., Kee, N.K., Zheng, J., Thalmann, N.M.: Design and development of a virtual dolphinarium for children with autism. IEEE Trans. Neural Syst. Rehabil. Eng.: Publ. IEEE Eng. Med. Biol. Soc. **21**(2), 208–217 (2013)
6. Ehrlich, J.A., Miller, J.R.: A virtual environment for teaching social skills: AViSSS. IEEE Comput. Graph. Appl. **29**(4), 10–16 (2009)
7. Cheng, Y., Chiang, H.C., Ye, J., Cheng, L.H.: Enhancing empathy instruction using a collaborative virtual learning environment for children with autistic spectrum conditions. Comput. Educ. **55**(4), 1449–1458 (2010)
8. Matson, J.L., Benavidez, D.A., Stabinsky Compton, L., Paclawskyj, T., Baglio, C.: Behavioral treatment of autistic persons: a review of research from 1980 to the present. Res. Dev. Disabil. **17**(6), 433–465 (1996)
9. Matson, J.L., Hattier, M.A., Belva, B.: Treating adaptive living skills of persons with autism using applied behavior analysis: a review. Res. Autism Spectr. Disord. **6**(1), 271–276 (2012)
10. Centers for Disease Control and Prevention.: Facts about autism spectrum disorders. Retrieved from http://www.cdc.gov/ncbddd/autism/data.html (2014)
11. Baron-Cohen, S., Leslie, A.M., Frith, U.: Does the autistic child have a "theory of mind"? Cognition **21**(1), 37–46 (1985)
12. Dichter, G.S.: Functional magnetic resonance imaging of autism spectrum disorders. Dialogues Clin. Neurosci. **14**(3), 319 (2012)
13. Pelphrey, K.A., Sasson, N.J., Reznick, J.S., Paul, G., Goldman, B.D., Piven, J.: Visual scanning of faces in autism. J. Autism Dev. Disord. **32**(4), 249–261 (2002)
14. Klin, A., Jones, W., Schultz, R., Volkmar, F., Cohen, D.: Visual fixation patterns during viewing of naturalistic social situations as predictors of social competence in individuals with autism. Arch. Gen. Psychiatry **59**(9), 809–816 (2002)
15. Von Hofsten, C., Uhlig, H., Adell, M., Kochukhova, O.: How children with autism look at events. Res. Autism Spectr. Disord. **3**(2), 556–569 (2009)
16. Chawarska, K., Shic, F.: Looking but not seeing: atypical visual scanning and recognition of faces in 2 and 4-year-old children with autism spectrum disorder. J. Autism Dev. Disord. **39**(12), 1663–1672 (2009)
17. Kolb, B.: Brain Plasticity and Behavior. Psychology Press, United Kingdom (2013)
18. Boraston, Z., Blakemore, S.J.: The application of eye-tracking technology in the study of autism. J. Physiol. **581**(3), 893–898 (2007)
19. Duchowski, A.: Eye Tracking Methodology: Theory and Practice, vol. 373. Springer, Berlin (2007)
20. Laganière, R.: OpenCV 2 Computer Vision Application Programming Cookbook: Over 50 Recipes to Master this Library of Programming Functions for Real-Time Computer Vision. Packt Publishing Ltd, Birmingham (2011)
21. Guestrin, E.D., Eizenman, E.: General theory of remote gaze estimation using the pupil center and corneal reflections. IEEE Trans. Biomed. Eng. **53**(6), 1124–1133 (2006)
22. Tobii EyeX—experience eye tracking in apps or games. http://tobii.com/en/eye-experience/eyex. Accessed 17 Oct 2014
23. Unity—Game Engine. http://unity3d.com/. Accessed 17 Oct 2014
24. Guillon, Q., Hadjikhani, N., Baduel, S., Rogé, B.: Visual social attention in autism spectrum disorder: insights from eye tracking studies. Neurosci. Biobehav. Rev. **42**, 279–297 (2014)

Printed in the United States
By Bookmasters